Literature and the Anthropocene

The Anthropocene has fundamentally changed the way we think about our relation to nonhuman life and to the planet. This book is the first to critically survey how the Anthropocene is enriching the study of literature and inspiring contemporary poetry and fiction. Engaging with topics such as genre, life, extinction, memory, infrastructure, energy, and the future, the book makes a compelling case for literature's unique contribution to contemporary environmental thought. It pays attention to literature's imaginative and narrative resources, and also to its appeal to the emotions and its relation to the material world. As the Anthropocene enjoins us to read the signals the planet is sending and to ponder the traces we leave on the Earth, it is also, this book argues, a literary problem.

Literature and the Anthropocene maps key debates and introduces the often difficult vocabulary for capturing the entanglement of human and nonhuman lives in an insightful way. Alternating between accessible discussions of prominent theories and concise readings of major works of Anthropocene literature, the book serves as an indispensable guide to this exciting new subfield for academics and students of literature and the environmental humanities.

Pieter Vermeulen is an associate professor of American and comparative literature at the University of Leuven, Belgium. He is the author of *Romanticism after the Holocaust* and *Contemporary Literature and the End of the Novel: Creature, Affect, Form.*

Literature and Contemporary Thought

Literature and Contemporary Thought is an interdisciplinary series providing new perspectives and cutting edge thought on the study of Literature and topics such as Animal Studies, Disability Studies and Digital Humanities. Each title includes chapters on:

- why the topic is relevant, interesting and important at this moment and how it relates to contemporary debates
- the background of and a brief introduction to the particular area of study the book is intended to cover
- when this area of study became relevant to literature, how the relationship between the two areas was initially perceived and how it evolved

Edited by Ursula Heise and Guillermina De Ferrari this series will be invaluable to students and academics alike as they approach the interdisciplinary study of Literature.

Available in this series:

Literature and Law
Mark Fortier

Literature and the World
Stefan Helgesson and Mads Rosendahl Thomsen

Literature and Materialism
Frédéric Neyrat

Literature and Social Media
Bronwen Thomas

Literature and the Anthropocene
Pieter Vermeulen

Literature and the Anthropocene

Pieter Vermeulen

Routledge
Taylor & Francis Group
LONDON AND NEW YORK

First published 2020
by Routledge
2 Park Square, Milton Park, Abingdon, Oxon OX14 4RN

and by Routledge
52 Vanderbilt Avenue, New York, NY 10017

Routledge is an imprint of the Taylor & Francis Group, an informa business

British Library Cataloguing-in-Publication Data
A catalogue record for this book is available from the British Library

Library of Congress Cataloging-in-Publication Data
A catalog record has been requested for this book

ISBN: 978-1-138-54371-3 (hbk)
ISBN: 978-1-138-54374-4 (pbk)
ISBN: 978-1-351-00542-5 (ebk)

Typeset in Sabon
by Taylor & Francis Books

Contents

Acknowledgments

This book concludes a five-year period of thinking and writing about the relation between literature and the Anthropocene. I want to thank the people who invited me to talk or write about this topic and encouraged me to develop my ideas: Brigitte Adriaensen, Lucy Bond, Holly Brown, Liliane Campos, Marco Caracciolo, Stef Craps, Rick Crownshaw, Ben De Bruyn, Kári Driscoll, Joanna Freer, Tom Idema, Suzanne Knittel, Yvonne Liebermann, Birgit Neumann, Jessica Rapson, Mads Rosendahl Thomsen, and Lars Saetre. In Leuven, I was fortunate to be able to discuss many ideas with my colleagues at the literature department, and I especially want to thank Lieven Ameel, Sascha Bru, Ortwin de Graef, and Tom Toremans for their companionship and advice. Tom Chadwick, Kahn Faassen, Reuben Martens, Ella Mingazova, and Ioannis Tsitsovits developed congenial doctoral projects from which I learned a lot. Many of the ideas in this book were beta tested on an excellent group of students in a 2018 masters course on contemporary American literature and the Anthropocene. Leanne Rae Darnbrough and Lisanne Meinen read most of the manuscript and provided invaluable feedback in the final stages. At home, Mats and Stine remind me every day that finding forms for planetary collapse is not a merely theoretical exercise, but a matter of survival. I couldn't wish for a better companion than Mirjam to spend the end of the world with. I dedicate the book to her.

Series Editors' Preface

Since the turn of the millennium, literary and cultural studies have been transformed less by new overarching theoretical paradigms than by the emergence of a multitude of innovative subfields. These emergent research areas explore the relationship between literature and new media technologies, seek to establish innovative bridges to disciplines ranging from medicine, cognitive science, social psychology to biology and ecology, and develop new quantitative or computer-based research methodologies. In the process, they rethink crucial concepts such as affect, indigeneity, gender, and postcolonialism and propose new perspectives on aesthetics, narrative, poetics, and visuality.

Literature and Contemporary Thought seeks to capture such research at the cutting edge of literary and cultural studies. The volumes in this series explore both how new approaches are reshaping literary criticism and theory, and how research in literary and cultural studies opens out to transform other disciplines and research areas. They seek to make new literary research available, intelligible, and usable to scholars and students across academic disciplines and to a broader public beyond the university interested in innovative approaches to art and culture across different historical periods and geographical regions.

Literature and Contemporary Thought highlights new kinds of scholarship in the literary and cultural humanities that are relevant and important to public debates, and seeks to translate their interdisciplinary analyses and theories into useful tools for such thought and discussion.

Ursula K. Heise and Guillermina De Ferrari

Introduction: Naming, Telling, Writing—The Anthropocene

Literature and the Naming of the Present

In the last decade, the notion of the Anthropocene has gained an almost viral popularity. After it emerged in the earth sciences, use of the term has proliferated in the art world, in environmental thought, increasingly also in literary and cultural studies, and, more recently, in public discourse. But if the term has many users, it has few enthusiastic fans, as many consider the term too abstract and too grandiose, too hip and too glib. As we will see, there are good reasons for this resistance. Still, that dissatisfaction is no reason to let go of the word (as if words could just be policed out of existence): disappointment is almost unavoidable in the face of the far-reaching environmental and social challenges that the Anthropocene is supposed to name. Naming the present is necessarily provisional and ongoing work, and it is a task that is inevitably speculative and imaginative. This is only one reason why literary and environmental thinking can enrich one another in coming to terms with the Anthropocene.

Take, as an example, Jeff VanderMeer's (2014a) *Annihilation*, the first novel in the *Southern Reach* trilogy. *Annihilation* tells the story of an expedition of four scientists sent to investigate the mysteries of Area X, a seemingly pristine region of coastline separated from the rest of the United States by an invisible border. Life in Area X is not merely wild, it is positively weird: on top of black bears, coyotes, and huge aquatic reptiles, there is also an undetermined "low, powerful moaning at dusk" (5), and the crew later stumbles upon "a vast biological entity that might or might not be terrestrial" (90). The status of this entity, which will later be called "the Crawler," is deeply uncertain: does it belong to the earth? Is it a cosmic force? Or a shared hallucination by the crew members? This radical ambivalence makes Area X "weird" in a more strictly literary sense. In literary history, the notion is customarily linked to the so-called “old weird” fiction of writers like H.P. Lovecraft, who wrote cosmic horror stories that, in Lovecraft’s own words, indulged a “breathless and unexplainable dread of outer, unknown forces” and evoked “the daemons of unplumbed space” (Lovecraft 2011, 1043). Like the old weird, the work of VanderMeer is invested in strange and disorienting entities and feelings. Still, and unlike the old weird, VanderMeer’s so-called “new” or “ecological” weird does not allow the human to survive the encounter with alien forces unscathed. In the *Southern Reach* trilogy, it is not clear whether the strange entity is an illusion, a creature, or “‘merely’ a machine” (191), but it is clear that human life will have to find ways to live together with it as the entity actively “creates out of our ecosystem a new world” (191). While the title of the first instalment in the trilogy evokes a feeling of cosmic horror (annihilation!), the final instalment (*Acceptance*) conveys a sense that altering the terms on which human and nonhuman lives coexist is something to welcome rather than resist.

Still, the trilogy’s embrace of the erosion of the borders between the human and the nonhuman is not unconditional. Many of the institutions and forms of life that we have constructed rely on a firm distinction between human and nonhuman entities. Our legal systems, for instance, assume that only human beings ever do things intentionally. If the organism in *Annihilation*, then, has an “intelligence … far different from our own” and displays “processes and aims [that] are utterly alien” (191), the novel invites us to consider the unfamiliar idea that a nonhuman thing can also have aims, plans, and designs of its own, and that human beings now have to share their agency with other kinds of forces. The narrator, a biologist, wonders what the nonhuman entity’s alien intelligence might mean “for the collection of cells and thought” that comprise her (172), when even the thoughts she thinks that she thinks might not be her own, but rather the effect of the organism “pulling these different impressions of itself from [her] mind and projecting them back at [her], as a form of camouflage” (179). It is not just that the boundary between human and nonhuman is an unstable one, it is also that the border seems to cut right through human communities, bodies, and even psyches. Human life, *Annihilation* suggests, is fundamentally *entangled* with nonhuman lives.

So how do we begin to understand those entanglements? How do we approach a reality in which traditional divisions no longer hold? What happens to the scientific disciplines through which we customarily study the world? In this new dispensation, the neat division between the sciences (which study nature in splendid isolation from human intervention) and the humanities (traditionally dedicated to the study of society and culture) becomes an obstacle in the effort to come to terms with a new reality in which the human and nonhuman realm are deeply entwined. Neither the sciences nor the humanities are particularly well equipped to capture that entanglement on their own. It will be one central argument of this book that literature is an important resource for approaching this uncharted territory, this intellectual Area X—and this is why I begin this book with a literary example, the first of many to follow. In VanderMeer's trilogy, the eponymous Southern Reach is a government agency commissioned to study and conceal Area X. It confronts Area X's challenge to existing disciplinary divisions by delegating an interdisciplinary team, consisting of a biologist, an anthropologist, a surveyor, and a psychologist (the linguist drops out at the last moment). At the end of the novel, only the biologist survives. Rather than an interdisciplinary sum of existing disciplines, it seems, we will need novel ways of apprehending the entanglement of human and nonhuman lives. In the first chapter of this book, I show how the interdisciplinary field of the environmental humanities has, in recent years, increasingly come to serve as a rubric for such innovative knowledge work. Because developments in this field resonate with exciting innovations in contemporary literature, this book will not only survey the interface of literary and environmental thinking, but also present and analyze a significant sample of the most pertinent literary engagements with the altered relation between the human and the planet.

At the beginning of the century, that new planetary condition received the by now familiar name of the Anthropocene (literally meaning "the new epoch of humans"). While the term had been used before by the biologist Eugene Stoermer, it became prominent when the Nobel Prize-winning atmospheric chemist Paul Crutzen declared in 2000 that the Holocene, the relatively stable and mild geological epoch in which human life had flourished for 11,700 year, was coming to an end, and that we needed a new name for a new epoch. The culprit: human life. In the past three centuries, Crutzen notes, human action has come to expand its reach over the whole planet, and its impact on the climate and on chemical and geological processes has become ever more intensive, which leads him to suggest the Anthropocene as the proper name for this new epoch in which humans have become a proper geological (rather than merely social or biological) agent akin to meteorites and volcanoes. Crutzen (2002) looks around the globe and sees a planet that has been thoroughly shaped by human intervention:

> About 30–50% of the planet's land surface is exploited by humans. Tropical rainforests disappear at a fast pace, releasing carbon dioxide and strongly increasing species extinction. Dam building and river diversion have become commonplace … Energy use has grown 16-fold during the twentieth century, causing 160 million tonnes of atmospheric sulphur dioxide emissions per year, more than twice the sum of its natural emissions. More nitrogen fertilizer is applied in agriculture than is fixed naturally in all terrestrial ecosystems; nitric oxide production by the burning of fossil fuel and biomass also overrides natural emissions.
>
> (23)

Crutzen's litany (which is about twice as long as the parts I quote here) makes the oversized impact of human action on the planet absolutely clear. At the same time, it is hard to imagine a reader capable of understanding all elements of it. To fully appreciate the significance of the facts that Crutzen enumerates, a reader would need to know demographics, atmospheric chemistry, climatology, engineering, and more. This is not something *Annihilation*'s interdisciplinary crew could pull off.

This book not only argues for the relevance of literature, but also of literary studies, for coming to terms with our Anthropocene condition. Crutzen's catalogue offers a good occasion to illustrate that relevance (and we will see in the third chapter on "Objects, Matters, Things" that the catalogue or the litany is a key stylistic feature in Anthropocene writing). If *Annhilation*'s crew had contained someone trained in the study of language, and especially literature (and remember it almost included a linguist), that someone could have suggested that an information dump that ranges across disciplines is less effective for the disinterested communication of facts than for making a particular affective impression on readers. Such an accumulation of facts, delivered in a monotone sequence of declarative sentences, not only aims to *inform* readers—it is as likely to *overwhelm* and *exhaust* them. Literary studies has the expertise to explain that effect. Crutzen's relentless and cheerless accumulation of worrying facts brings readers to the limits of their capacity to process information. In that sense, it has something to do with the romantic aesthetic of the sublime, which aimed to inspire awe and terror, yet it deviates from it in providing the particular combination of *astonishment* and *boredom* that cultural critic Sianne Ngai has identified as "stuplimity." The traditional sublime staged "the instantaneous or sudden defeat of comprehension" in the face of the infinite and elemental (not unlike Lovecraft's cosmic horror); stuplimity, for its part, rather operates through "an extended duration of consecutive fatigues" (Ngai 2007, 19). The dispiriting effect of the stuplime consists in a combination "of boredom and astonishment, of what 'dulls' with what 'irritates' or agitates, of excessive excitation with extreme desensitization or fatigue" (16). As Ngai explains, such a creeping fatigue does not inspire action or a desire to escape from the systems in which we are trapped. Her category helps us understand the awkward and disorienting effect that Crutzen's catalogue has on the reader: half-informed, overwhelmed, aware that things are bad and need to change, ... but then what?

Descriptions of the Anthropocene, even seemingly sober and descriptive ones like Crutzen's, are never simple statements of fact. We tend to think that a straightforward sum of incontrovertible scientific observations adds up to a watertight conclusion—in this case, the realization that the moderations of the Holocene are over, and that human action has shifted the planet to a new epoch, an epoch we can name the Anthropocene. Yet the process of *naming* a new reality is considerably more complicated, for at least three reasons. First, the facts that point to the reality of the Anthropocene are in no way simple empirical observations: measuring global sulphur dioxide emissions and comparing those numbers to past emissions and to future reserves, for instance, relies on the harvesting of global data on a scale that can only be processed by vast computational infrastructures—what Paul Edwards (2000) has called the "vast machine" of data models, simulation models, and reanalysis models (xv). Without observation instruments, algorithms, and the availability of vast computer processing capacities, there would simply be no data, and no scientific facts about the changing earth system. For climate change skeptics and deniers, the obvious conclusion is that these data are fabricated, doctored, and distorted; for people more familiar with the procedures of science, this is simply how facts are established. Edwards makes it clear that observation and modeling have generated a clear consensus about the reality of climate change since the 1990s (7).

But how do we translate such a robust scientific consensus into a new name? Like all speech acts, that of baptizing a geological epoch relies on a number of institutional conditions (and this is a second complication). These institutional realities are quite sobering. Since 2009, an interdisciplinary Anthropocene Working Group (AWG) has been mandated by the International Commission on Stratigraphy's (ICS) Subcommission on Quaternary Stratigraphy (SQS) to assess the viability of the term as a geochronological unit. The group's 2017 "Summary of Evidence and Interim Recommendations" makes evident the entanglement of scientific observation and bureaucratic process. The group "holds the Anthropocene to be stratigraphically real," as there is a sufficient number of stratigraphic signals that show the earth system moving away from Holocene values (Zalasiewicz et al. 2017, 55). The group's conclusions, we read, were not reached by consensus, but by an "informal and non-binding" email ballot (58). After eight years, the working group points out that a lot of work remains to be done: the selection of sites for further analysis, a full description of relevant signals, the preparation of a formal proposal to the SQS, endorsement by the SQS, referral to the ICS, and, in case of a favorable vote, ratification by the Executive Committee of the IUGS (59). We are, to say the least, not quite there.

Except, of course, that we are. As the world is witnessing wildfires at the Arctic and Brazil's accelerating destruction of the Amazon, the whole ratification process seems somewhat beside the point. At the end of the Working Group's paper, the authors note the irrelevance of the bureaucratic steps they have just patiently enumerated: "Whichever way this particular process ends, it is clear that human beings are now operating as a major geological agent at the planetary scale, and that their activities ... have imprinted an indelible mark on the planet" (59). Crutzen (2002) himself similarly registers some impatience with terminological disputes and ends his 2002 article by reminding his audience of the challenges ahead: "A daunting task lies ahead for scientists and engineers ... This will require appropriate human behaviour" (23). As Crutzen's fairly provincial appeal to only "scientists and engineers" already suggests, he believes in the viability of "large-scale geo-engineering projects," but he admits that this preference is not based on scientific fact but remains a matter of conjecture: "At this stage," he concludes, "we are still largely treading on *terra incognita*" (23). A third reason the Anthropocene is a contested name, then, has to do with the inevitable slippage between painstakingly established scientific facts and deeply-felt value commitments, between patient description and an urgent need to act.

The fact that the production of data relies on models and infrastructures; the slow ratification process by the geological community, which makes it easy to dismiss artistic, literary, or humanities-based engagements with the notion of the Anthropocene as premature (even if, in reality, they might already be too late); and finally the fact that the notion registers less "a matter of fact" than what Bruno Latour (2004), an important voice in humanistic reflections on the Anthropocene, has called "a matter of concern"—all three factors make the Anthropocene a name that will never be fully adequate, and that can only ever be a *misnomer* for a complex and open-ended cluster of observations, procedures, affects, commitments, and values. The notion has few explicit advocates, and people using the term are often quick to point out their own reservations about it. But if no one is exactly *for* the Anthropocene (with the possible exception of Diane Ackerman [2014] in her book *The Human Age*), it has undeniably been productive as a term of debate. The Anthropocene is something one can be *against* (as in the titles of works by Daniel Hartley [2015] and T.J. Demos [2017]), and it is a name that has invited many alternative coinages that claim to better capture current realities and challenges. Steve Mentz (2019) has identified no less than 24 'cenes that have been coined in what he whimsically calls "the Neologicismcene"—underlining that "the new epoch of humans" is also the epoch in which humans can't stop inventing new names for their predicament, if only because no one name precisely nails it. In the next section, we will look at notions such as the Chthulucene, the Capitalocene, and the Plantationocene. These terms in different ways argue that the Anthropocene's invocation of a human collective overlooks substantial differences between different human communities, and fails to convey that some (typically privileged) constituencies bear much more responsibility for the ongoing planetary crisis than the (often disadvantaged) groups that suffer from it most directly. The term, for the proponents of these alternative names, is not only a misnomer, but also serves as a kind of disingenuous *disclaimer* that dissolves accountability.

The Anthropocene as misnomer and disclaimer: perhaps such linguistic infelicity is unavoidable when we are inhabiting what Crutzen calls "*terra incognita*," an earth we no longer know. Our disciplinary traditions, whether in the humanities or the sciences, have not prepared us for the novel constellations of the human and the nonhuman. Let's turn to VanderMeer's *Annihilation* again. Here, the *terra incognita* is called Area X, and is not situated in the void, but in the debris of human civilization, amidst "eerie signs of human habitation" (5). In this realm, names don't function as they do in more hospitable environments, and none of the characters are referred to by their names, only by their disciplinary identities (which, as we saw before, also fall short of understanding the weirdness of the area). As the biologist notes: "we were always strongly discouraged from using names: We were meant to be focused on our purpose, and 'anything personal should be left behind.' Names belonged to where we had come from, not to who we were while embedded in Area X" (9).

"Embedded in Area X": this pretty much describes the way human life now has to find a way to cohabitate with the nonhuman forces with which it is irrevocably entangled. This is a situation, *Annihilation* notes, in which customary names don't belong. It is a situation in which language, rather, needs to find ways to test and tune an unsettling and untried constellation of things, forces, and affects. This is precisely what VanderMeer's atmospheric, patient, and almost ambient mode of describing Area X achieves. If I have drawn on a literary work to introduce the multifarious complexities that beset the name "Anthropocene," it is from the conviction (which I share with VanderMeer and many of the authors I turn to in the rest of the book) that literature's imaginative, narrative, affective, and reflexive resources have a role to play in updating the knowledge infrastructures we need to reconnoiter new intellectual territories. If Crutzen believes that "scientists and engineers" will help us settle these new territories, this book shows how literature and literary studies can help the broader project of the environmental humanities unsettle such engrained but obsolete disciplinary certainties.

The inadequacy of the name of the Anthropocene is one reason why this book sticks to it. The name is less a rigid designator with a stable referent (for all the reasons discussed above) than a rubric that has, since the beginning of the century, increasingly come to cluster concerns over the human impact on the planet. The term has been undeniably productive as a catalyst for ecological concerns, and for discourses and practices through which human anxieties and aspirations are articulated. The term has generated fewer defenses than arguments against it or in favor of alternatives for it, but such arguments end up demonstrating the usefulness of the term in fueling argumentation, reflection, and debate about crucial aspects of the genealogies, challenges, and prospects that make up the present. The Anthropocene, in other words, becomes useful if we accept that is inevitably a misnomer. It covers a makeshift assemblage of discourses, terms, protocols, and experiments that never fully hit home—if only because home turns out to be a weirder place than we remember. For Bruno Latour (2017), the reason to stick to the name is that it makes it possible to "stay with the trouble" (100n77). Latour borrows this phrase from Donna Haraway, who, even though she rejects the term, calls for an attempt "to be truly present"; for Haraway (2016), the challenge is to refuse to transcend the challenges of the present, and to inhabit the "now" as "mortal critters [Haraway's term for 'microbes, plants, animals, humans and nonhumans'] entwined in myriad unfinished configurations of places, times, matters, and meanings" (1, 169n1). As I show in the rest of this book, literature and the study of literature are two vital resources for learning to inhabit the present—for the hard work of what Haraway calls "getting on together" (10).

The Anthropocene and Its Others

Except in the rarified realm of stratigraphy, the Anthropocene has no robust real-world referent, but it has managed to assemble a number of phenomena that provide evidence for the far-reaching consequences of human action. The most obvious one is climate change, which entered public consciousness in the 1980s (well before the notion of the Anthropocene began its career), often under the name of global warming. While the Anthropocene is popularly still tied to anthropogenic climate change, the term also covers processes such as ocean acidification (which is a consequence of increasing carbon dioxide levels), global population growth, resource depletion, massive species extinction, and ecosystem simplification more generally. Cumulatively, these phenomena point to the *unsustainability* of the human exploitation of the planet. Our species uses about one third of the continental biomass available every year for its own sustenance; 97 per cent of the biomass of land vertebrates consists of humans and their domestic animals (Bonneuil and Fressoz 2016, 7–9). And the planet is not only full of human life, it is also full of the nonhuman stuff that is needed to support that life. Jedediah Purdy (2019) reports that "the material habitat that humans have created for themselves in the form of roads, cities, rural housing, the active soil in cropland, and so forth" is estimated "at thirty trillion tons, five orders of magnitude greater than the weight of the human beings that it sustains" (21–22). The Anthropocene expresses the fact that human designs have thoroughly colonized the planet to the point that no margins or sites of refuge for different ways of being or even biological diversity remain.

The Anthropocene presents a thoroughly humanized earth, but this does not mean that earth is under human control. The problem is that the different manifestations of the Anthropocene are neither linear nor localized, neither reversible nor containable. Developments like population increases, intensive farming, and the unlocking of energy resources have at one time undeniably served to increase human welfare, and in some cases still do, but they have set off ecological and chemical processes that interlock with one another in ways that destabilize the earth system as a whole. In the Anthropocene, the hydrosphere (the watery part of the earth), the atmosphere (gases), the cryosphere (the planet's ice mass), the lithosphere (the solid parts), but also the economic and energy systems that tap into these and the social and symbolic orders that we have developed to inhabit them have begun to interact in ways that are less predictable than ever.

Take, for example, the effect of rising temperatures on Arctic ice melt: disappearing ice will result in warmer and expanding oceans, rising sea-levels, threats to coastal constituencies, patterns of geographical dislocation and mass migration, political instability, among others. If these effects are increasingly speculative (the first few are decidedly not), they are also far from unlikely. Another feature of Arctic thaw is that the process is self-accelerating: once sea ice disappears, less of the sun's heat will be reflected and more will be absorbed by the planet, which will accelerate the warming that propels the melting in a positive feedback loop. The instability of the earth system is often expressed through the notion of "tipping points": moments when accumulated changes in degree *suddenly and irreversibly* shift to changes in kind. Literary critic Timothy Clark (2015) has referred to the Anthropocene as a "threshold concept": a "necessarily vague but insidious border" (48) beyond which the predictions, practices, and habits on which we rely become inoperative. This is the terrain Paul Crutzen referred to as *terra incognita*.

Clark emphasizes that the Anthropocene is also a threshold concept for the ways we make sense of the world: it names a brink where "modes of thinking and practices that were once self-evidently adequate, progressive, or merely innocuous become, in this emerging and counter-intuitive context, even latently destructive" (21). For Claire Colebrook (2012) too, the Anthropocene calls for new concepts, imaginaries, and vocabularies, as it provides "a threshold at which all 'our' concepts of horizon, milieu, ethos and polity are voided" (188). Even "the terms of our ethical vocabulary—justice, fairness, respect, forgiveness, hospitality or virtue" (185)—are no longer self-evident. Indeed, what does hospitality mean when one species (ours) has so clearly crowded out its hosts? How can we describe behavior as virtuous when it may have unpredictable long-term outcomes that harm distant others? And, to shift from ethical to political challenges, what is the role of nations when responses to global destabilization seem to require supranational forms of governance *beyond* the nation? How viable are forms of "eco-cosmopolitanism" that imagine connections across national borders (Heise 2008) or multispecies assemblages that look for interactions across species borders (van Dooren et al. 2016)? These questions are new and urgent; they are as hard as they are unavoidable.

Most criticisms of the Anthropocene believe the term not only generates these questions, but also directs us to incorrect or insufficient answers. In the rest of this section, I highlight four such criticisms, and introduce the (often playful) terminological alternatives that critics have proposed. I consider the criticism that the notion misrepresents the central role of capitalism in the exploitation of the natural world; that it unjustifiably unifies human constituencies at the abstract level of the species; that it overlooks the implication of nonhuman actors in changing planetary constellations while it privileges technological fixes and succumbs to technological determinism; and that it overplays the uniqueness of the current crisis and fails to locate present concerns in a longer history of ecological awareness. All of these points are relevant, and they underline that the Anthropocene is more useful as a catalyst for debate than as a fully adequate name for the present. In this spirit, we can welcome notions such as the Capitalocene, the Plantationocene, the Chthulucene, or even the Oliganthropocene and the Homogenocene as helpful additions to the ongoing debate over the challenges we face.

The Anthropocene Misrepresents the Role of Capitalism (Capitalocene, Plantationocene)

When Paul Crutzen coined the notion of the Anthropocene, he proposed the late eighteenth century as its starting point. This date is not uncontested: as we will see below, the Anthropocene Working Group proposes the post-Second World War period as a more appropriate marker. Crutzen wanted to foreground the role of the Industrial Revolution in the alteration of the earth system. On this account, James Watt's invention of the double action steam engine in 1784 kick-started the transition to coal fuel that powered the Industrial Revolution that in its turn triggered the relentless rise of atmospheric carbon dioxide levels. This perspective privileges technological innovation over capitalist dynamics. As Andreas Malm (2016) has shown, fossil fuel is not magically more efficient, abundant, or reliable than, for instance, hydraulic power, but it is easier to make a profit from it: fossil fuel extraction can easily be located in areas where labor is cheap, and it allows for competition (rather than requiring intense collaboration, as hydropower does) between entrepreneurs.

This is only one example of how far-reaching ecological changes emerge as a result of capitalist forces rather than technological breakthroughs. Proponents of the term *Capitalocene* argue that Crutzen's "official" narrative is not only too focused on technology, but also on Western modernity, especially on the British Industrial Revolution, and fails to factor in that capitalism has reorganized the global ecological system in an encompassing and intensive way for centuries. The viability of investments in steam energy, for instance, depended on the availability of the transatlantic slave trade, which provided cheap labor; of large and violently depopulated territories in the New World; and of a broad global demand for inexpensive cotton cloth (Malm and Hornborg 2014, 63–64). Capitalist trade networks, in other words, had to be in place for Watt's steam engine to have the effect it had. The immensely profitable cotton monoculture in the Caribbean was only possible because slaves were fed by agricultural products from North America and cod from Newfoundland, while Britain could only dedicate its own soil to the production of grain, wood, and animal fodder by "outsourcing" the cultivation of cotton fiber to its colonies (Bonneuil and Fressoz 2016, 232–33). The notion of the Capitalocene wants to highlight how global trade networks and class and colonial relations fundamentally shaped the planet. It reminds us that capitalism, ever since its emergence in the fifteenth century, has revolutionalized the landscape and has been an "environment-making" force. In Jason Moore's (2017) words, the notion situates the rise of capitalism "historically and geographically, within the web of life" (608–609).

Proponents of the Capitalocene are not unaware that the human impact on the make-up of the planet is not exclusive to capitalism. The ecological record of twentieth-century communist states is notoriously dismal, and today not-quite-capitalist China is the world's largest source of carbon emissions (in total, not per capita). Paleoclimatologist William Ruddiman (2003) has proposed the so-called "Early Anthropocene

Hypothesis": he argues that early agriculture, starting well before capitalism some 8,000 years ago, initiated emissions of greenhouse gasses that affected climate developments. Still, foregrounding capitalist dynamics is helpful for comprehending the massive *scale* of human impact (and I turn to the crucial issue of scale in my third chapter), and it has the potential to politicize talk about the environmental crisis. Capitalist exploitation has always entailed the redesigning of landscapes (think of the imposition of cotton monocultures) and the circulation of bodies (think of the slave trade).

The notion of the *Plantationocene* (launched tongue-in-cheek during a 2014 roundtable) nominates the plantation as the paradigmatic historical site where capitalism wreaks its destructive work (Haraway et al. 2016, 556). Situated at the intersection of forcibly displaced labor, long-distance financial investment, and intensive cultivation of the soil, the plantation is a systematic practice of relocation that initiates a major upheaval in the relations between humans, animals, plants, and other organisms. It conveniently allowed privileged constituencies to treat violence to people and the environment as a distant and invisible concern. The slave plantation system, in Donna Haraway's (2016) words, "was the model and motor for the carbon-greedy machine-based factory system," and it "continues with ever greater ferocity in globalized factory meat production, monocrop agribusiness, and immense substitutions of crops like oil palm for multispecies forests and their products" (206n5). The notion of the *Homogenocene*, proposed by the entomologist Michael Samways in 1999, captures this sense of a planet in which biological and cultural life have increasingly become the same everywhere (Mann 2011, 23). Like the Capitalocene, the Plantationocene reminds us that global change is less the result of technological breakthroughs than of structural processes that move human bodies and other biomass around the world in the name of profit.

The Anthropocene Gives Us an Insufficiently Differentiated Humanity (Oliganthropocene)

These accounts make clear that different constituencies are differently implicated in processes of planetary change, and that the notion of the Anthropocene levels out these differences through the overly abstract notion of the *anthropos*, the human. Analytically, it is less than useful to group, for instance, chattel slaves and long-distance investors as members of the same species. Not only do affluent Western consumers with vast carbon footprints bear a disproportionally larger responsibility for environmental deterioration than destitute peasants in the Global South, the fallout of those changes also affects different constituencies in vastly dissimilar ways: the impact of Hurricane Katrina on black and white neighborhoods in New Orleans or the threat of rising sea levels in a rich country like the Netherlands and an impoverished one like Bangladesh illustrate the differentiated vulnerability of distinct groups. This gives the lie to the facile claim that in a climate changed world, there are no lifeboats. As Malm and Hornborg (2014) write, "[f]or the foreseeable future … there *will* be lifeboats for the rich and privileged" (66). Today, such lifeboats often take the shape of a luxury doomsday bunker or even of a rocket, as in SpaceX CEO Elon Musk's much-hyped ambition to colonize Mars in order to, in Musk's own words, "safeguard the existence of humanity" (Assis 2015). The vaguely humanitarian rhetoric obscures that at best, planetary relocation can accommodate an infinitesimally small segment of the human population while it will abandon billions to a heating and drowning earth. In such scenarios, the pious invocation of the species transparently serves as an alibi *not* to address the degradation of the lifeworld. This state of affairs has led geographer Erik Swyngedouw to propose the term *Oliganthropocene* to accentuate the vast inequalities in the responsibility for and the response to environmental threats (Bonneuil and Fressoz 2016, 71).

Invocations of something as grandiose as the species need not be cynical or sinister. Historian Dipesh Chakrabarty, arguably the most influential theorist of the Anthropocene in the humanities, maintains that the upscaling of human life to the necessarily abstract level of the species holds a particular political promise. For Chakrabarty (2009), while it may be impossible to "experience ourselves as species" (220), and while our species identity thus necessarily remains an imaginative and speculative affair, this very abstractness holds room for the imagination of a better future that is not disabled by lingering legacies of injustice and inequality. Species, Chakrabarty surmises, "may indeed be the name of a placeholder for an emergent, new universal history of humans that flashes up in the moment of the danger that is climate change" (221).

One problem with the "shared sense of a catastrophe" (222) on which this imagining of human collectivity depends is that it obscures, in Rob Nixon's (2014) words, vastly "unequal human agency, unequal human impacts, and unequal human vulnerabilities." Nixon's own contributions to debates over the Anthropocene have advocated a disciplined focus on the storytelling practices and the forms of activism that make up an often overlooked "environmentalism of the poor." For Nixon (2011), the non-spectacular, attritional deterioration to which the global poor are exposed is hard to talk about and to make visible: such "slow violence" (his coinage), a violence of delayed destruction that is dispersed across time and space, does not really lend itself to the spectacle-driven templates of contemporary media culture, or even to the drama-dependent formats of literary storytelling (2). More attention to issues of environmental justice can enrich, nuance, and correct the all too sweeping talk about species that the notion of the Anthropocene invites.

The Anthropocene Fails to Factor in Nonhuman Agents (Chthulucene)

As the Anthropocene puts forward a robust and internally undifferentiated *anthropos*, it promotes the human as a supposedly unique and exceptional agent. The human is imagined to be not just the *cause* of current crises (the previous two subsections already pointed up the problems with that account), but also their *victim* and, most problematically, their heroic *solution*. Such a sustained emphasis on human autonomy forgets that environmental debasement has spun out of human control. This does emphatically not mean that all we can do is sit back and enjoy the show of our own undoing: affirming human responsibility and agency is crucial, yet so is insisting on the distinction between agency (which is a feature of human, nonhuman, and technological agents alike) and autonomy (which is an illusion that the Anthropocene forces us to abandon). Bruno Latour (2017) underlines that being a subject, and acting in a concerted and intentional way, does *not* mean "acting in an autonomous fashion": instead, "it means *sharing* agency with other subjects that have also lost their autonomy" (62). In a similar vein, Derek Woods (2014) has remarked that the subject of the Anthropocene is *not* the human species, but rather "the sum of terraforming assemblages composed of humans, nonhuman species, and technics" (134). The processes that are changing the ways we inhabit the earth are as much about the disempowerment as about the responsibility of human agents.

Official Anthropocene discourse does not see technology as a source of human disempowerment: technological devices are seen as prostheses that extend human power rather than as forces that also limit it. This conception feeds technocratic fantasies that we can engineer our way out of the crisis. For Crutzen, for instance, the notion of the Anthropocene highlights "the immense power of our intellect and our creativity, and the opportunities they offer for shaping the future" (Crutzen and Schwägerl 2011). In this upbeat discourse, capitalist markets are counted on to foster a "green economy" that will generate and promote sustainable solutions (Demos 2017). In the 2015 "Ecomodernist Manifesto," a group of eighteen scientists and activists cheerfully commit to the possibility of "decoupl[ing] human well-being from environmental destruction" through "a sustained commitment to technological progress" (Asafu-Asjaye et al. 2015, 29–31). Through technological developments, human action can transcend its environmental limitations and enter "a good, or even great, Anthropocene" (6). Limiting economic and technological development is unwelcome, as accelerated technological progress will make it possible to let natural territories "re-wild and re-green" (15). This scenario relies on a *denial* of shared agency; it celebrates human autonomy as a power that will release the nonhuman world from its entanglements with us. For ecomodernists, the technological contrivances we design allow us to transcend our ties to the planet and grant nature what we decide it deserves.

To convey human life's implication in the earth system, Bruno Latour introduces the terminological distinction between "the Humans" and what he calls "the Earthbound." The Humans believe we simply need an ecomodernist, technocratic update of the dispositions that successfully guided the species through the Holocene; Latour's (2017) coalition of the Earthbound, in contrast, accepts that the terms of our dwelling on the planet need to be renegotiated (248). The Earthbound are not at war with the planet, but they need to be prepared, Latour believes, to declare war on the Humans. What is at stake in that war are the imaginaries and practices through which we inhabit the unstable earth to which we are bound. What is needed, for Latour, is a promiscuous recombination of religious, scientific, artistic, and governmental practices. The Earthbound are both inevitably attached (*bound*) to the earth (or what Latour, borrowing from the geophysicist James Lovelock and the microbiologist Lynn Margulis, prefers to call the anti-systemic assemblage of "Gaia"), as well as heading (*bound*) for a different relation to the planet. Latour's imagined collective is by definition *not* the whole of humanity (this is different from Chakrabarty's species thinking); he underscores that a political struggle must be fought between opposing constituencies. The insistence on struggle allows Latour to emphasize human responsibility and agency *without* fantasizing about an abstract global collective.

Latour's intervention arguably stays too close to traditional human-centered perspectives. This is the point of multispecies thinker Donna Haraway, whose notion of the *Chthulucene* may be the most radical proposal to affirm the role of different life forms in the changing life-world. Against Latour's confrontational stance, Haraway (2017) calls for "compositionist practices" in which the Earthbound "entangle with the ongoing, snaky, unheroic, tentacular, dreadful ones, the ones which/who craft material-semiotic netbags [that are] of great use in bringing home and sharing the means of living and dying well" (40–43). The notion of the Chthulucene has nothing to do with H.P. Lovecraft's cosmic monster Cthulhu: as I noted in the previous section, most productive literary and theoretical engagements with the destabilized present complicate the allure of cosmic horror. Instead, it replaces the Greek *anthropos* with echoes of the Greek *khthonios*, which means "of the earth" (173–74n4). Haraway's earthbound multispecies assemblage avoids "a comic faith in technofixes" (3) and is instead committed to "staying with the trouble," to live, "somehow, on a vulnerable and wounded earth" (10). Haraway's temperamentally playful, intellectually promiscuous, and formally experimental intervention shifts the terms of debate away from high scientific seriousness to an almost comic exuberance: it advocates an "earthly worlding [that] is thoroughly terran, muddled, and mortal—and at stake now" (55).

The Anthropocene is Historically and Culturally Myopic

The "now" in which life is at stake points to the curious temporality of the Anthropocene, which will be an organizing concern of this book. The Anthropocene has burst upon the scene in the new millennium as if we *suddenly* realized that we have *always* been enmeshed with nonhuman agents. Suddenness and shock clash with the almost unimaginable slowness of geological and evolutionary time; urgency meets immobility. The exponential rise of Anthropocene talk makes it tempting to think of contemporary ecological awareness as unprecedented. Kate Marshall (2015) has baptized the current moment "the Anthropocene's reflexive phase" (11), a phrase that distinguishes our current hyperawareness from a period in which our actions unreflexively, almost unconsciously, eroded their geological and biological basis. Such a micro-scenario in which the present is endowed with a higher level of reflexivity than the past is all too familiar: it repeats the move in which modernity relegated the past to the status of dark and premodern Middle Ages. It also recalls sociologist Ulrich Beck's thesis of a second modernity or a reflexive modernization, in which society manages the unintended consequences of a reckless first modernity. In the context of the Anthropocene, positing such a sequence invites the idea that human reason and prowess have finally come to their senses and will engineer a way out of the crisis.

This two-step scenario is historically inaccurate. Christophe Bonneuil and Jean-Baptiste Fressoz's 2016 book *The Shock of the Anthropocene* is a magisterial riposte to such a "story of an awakening" (73). As they show, the period between 1770 and 1830, where the official narrative situates the start of the Anthropocene, was marked by a clear awareness of the tenuous interdependence of nature and society (76). Ecological awareness was not unavailable so much as it was actively marginalized. For Bonneuil and Fressoz, a facile narrative of reflexive awakening perpetuates that marginalization and distorts the prehistory of our current environmental concerns. The present is *not* blessed with a unique insight into the interrelations of human and nonhuman processes; it is, instead, "the culminating point of a history of destructions" (171). *The Shock of the Anthropocene*'s revision of the present shows that Western modernity cannot plead ignorance about its destruction of the planet: "our ancestors destroyed environments in full awareness of what they were doing" (196). This raises important questions about the role of knowledge in the human (lack of) response to environmental degradation. Is it a matter of outright *denial* of the facts? Or is the key the spread of *uncertainty* by what Naomi Oreskes and Erik Conway (2010) have called "merchants of doubt"? Or is it a massive case of *disavowal*, a situation in which people *know* the facts but yet find themselves unable to *act* on that knowledge ("I know very well … but all the same").

By robbing the past of its innocence and the present of its claim to intellectual superiority, such a historical perspective connects to Latour's intervention by affirming the ineluctability of struggle. Environmental awareness from the past can then be mobilized as part of an archive of resistance against planetary deterioration. Take, for instance, a literary classic like Mary Shelley's *Frankenstein* (1818). Traditionally read as a relay for anxieties over technological progress and science turning against the human life that created them, the novel also bears the imprint of Shelley's experience of the rural poverty and pervasive stormy weather that overcame Europe in 1816, the "Year without Summer." These disturbances were a consequence of the eruption of the Indonesian volcano Mount Tambora, a geological event that resonated around the globe (Higgins 2017; Wood 2014). Read in this light, the novel becomes a testimony to the tenuous interlinking of human lives and their environments and constitutes an early awareness of creaturely vulnerability that still speaks to the present.

Nor is it only the past that populates such an archive of awareness, as there are also founts of Indigenous ecological knowledge that remain largely unexplored. As I investigate in more detail in the fourth chapter ("Dominations"), for many Indigenous groups, the wholesale destruction of lifeworlds is nothing new but a recurring feature of life since the onset of the Western colonization of the world. For centuries, Indigenous cultures have been responding to the kind of environmental destruction that Western constituencies are only beginning to contemplate. As Kyle Powys Whyte (2017) notes, Anishinaabek (a group of related Indigenous people in Canada and the USA) are already living the dystopian future more privileged audiences only see on their film screens, as "settler colonial campaigns in the Great Lakes region have already depleted, degraded, or irreversibly damaged the ecosystems, plants, and animals that our ancestors had local living relationships with for hundreds of years" (207). The lived reality of resource extraction and environmental racism informs, for instance, a novel like Leslie Marmon Silko's *Almanac of the Dead* (1991), which imagines a cross-ethnic alliance against settler colonialism and environmental deprivation that is even more acute now than it was when the novel was published almost three decades ago (Streeby 2018, 52–58). If the Anthropocene can overcome its historical and cultural myopia, there is, unfortunately, a lot to learn from these archives.

Literature and the Anthropocene: Four Affordances

The short examples of *Almanac of the Dead* and *Frankenstein* point to the focus of this book: the interface of literature and the Anthropocene. They show how insights into the entanglement of nature and society and of life and geology make it possible to revisit literary history and read *Moby-Dick*, for instance, as a dramatization of the human's destructive mania in the face of a willful nature, or appreciate the extent to which Jack Kerouac's *On the Road* powers its iconic account of American freedom and transcontinental mobility through the obscene expenditure of fossil fuels. An Anthropocene perspective attunes us to elements that we have come to take for granted in the history of literature and art. Take, for instance, a painting like

Claude Monet's widely reproduced and taught *Impression: Sunrise* from 1873, which presents the color effects incumbent on a sunrise in the port of Le Havre. As Nicholas Mirzoeff (2014) has shown, many of the color effects we have learned to admire as the hallmarks of painterly genius reflect the coal smoke that the factory chimneys (almost obscured in the painting by these effects) blast into the atmosphere. The cultural celebration of impressionism, a movement that takes its name from this very painting, in a sense naturalizes as much as it reveals anthropogenic climate change. Yet literature and art do more than naturalize: if we look at the depiction of "fog and mud" and "smoke and soot" in the opening pages of Dickens's *Bleak House*, for instance, we can see Dickens use the London smog, an emblem of the interconnectedness of natural and social activity, to provide a "climate model" that depicts London, in Jesse Oak Taylor's (2016) words, "in climatic time, within which glaciers rise and recede, species evolve and go extinct, and civilizations decline and fall" (34).

This shows that literature is as much a repository of climate awareness as a resource for thinking about the entanglements between human and nonhuman actors. Dickens's literary climate-modeling, for instance, allows *Bleak House* to anticipate "more recent arguments for environmental justice, pointing out that the impacts of toxicity and ecological devastation are borne disproportionately by the poor" (J. Taylor 2016, 37). This is exemplified in the novel by the crossing-sweep Jo, who dies from his exposure to London's nasty climate. Rereading literary history from an Anthropocene perspective is contiguous with Nixon's interest in the environmentalism of the poor as well as with Bonneuil and Fressoz's recovery of half-forgotten instances of historical environmental awareness. In the rest of this section, I briefly discuss four features that qualify literature, like the arts more generally, as a valuable resource for confronting the Anthropocene: the centrality of *narrative* as a meaning-making device; literature's *affective* affordances as an aesthetic construct; its license to *imagine* possible scenarios; and finally, and less intuitively, literature's constitutive engagement with questions of writing, inscription, and action, which acquires a peculiar relevance in the Anthropocene.

Anthropocene Narrative

The intimate relation between the Anthropocene and narrative is fairly obvious: depending on the assumed starting point, theories of the

Anthropocene present so many different historical narratives with their own protagonists and plots. Four such proposals have received quite some airplay. If, like Paul Crutzen, we assume the Anthropocene to have started with James Watt's invention of the double action steam engine and the unfolding of the Industrial Revolution, the narrative that emerges is that of the human as an inventor and entrepreneur in a plot of capitalist expansion. If, like the AWG, we date the start to the so-called "Great Acceleration" after the Second World War (and this version is on its way to general acceptance), we are telling the story of an expansive consumer capitalism that spans the globe. Another suggestion is that by geographers Lewis and Maslin (2015), who propose to locate the start of the epoch in 1610, with the European conquest of the Americas and the onset of colonialism and global trade. This proposal entails a story in which the human protagonist features as a colonizer in an imperialist plot. Or take, finally, the Ruddiman hypothesis, which holds that pre-industrial agriculture already some 8,000 years ago emitted enough greenhouse gases to stall a new glacial episode and contributed to the moderation and stability of the Holocene. On this account, the particular kind of human that impacts the climate is a noble farmer, and the plot we are presented with is that of human settlement, not that of colonization, capitalist exploitation, or global conquest. In all of these versions, scientific insights into the geological impact of human life are mobilized to give narrative shape to the past and present. And as narratives inevitably encode particular forms of agency and obscure other ones, it matters what kind of stories we tell about the world.

Narrative also has an analytical function: it imposes patterns and order on an unruly reality, such as the reality of a planet spinning out of human control—a sense of control that was illusory to begin with. Increasingly, respectable scientists have turned to literary narrative to convey a coherent message about the disorderly present. In *The Collapse of Western Civilization: A View from the Future* (2014), historians of science Naomi Oreskes and Erik Conway adopt the voice of a Chinese historian living in 2393 and looking back on the fall of Western civilization three centuries earlier. The book is not interested in a science fictional account of the future, as it spends most of its energy developing a scientifically informed and dispassionately recounted story of the present and a probable assessment of the near future. The future perspective serves as a narrative-generating device, and the narrative in its turn serves an analytical goal: understanding how we "failed to act on robust information about climate change" because of science's "excessively stringent standard for accepting claims of any kind" and "an ideological fixation on 'free' markets" (Oreskes and Conway 2014, ix–x). The narrative thrust brings the scientific facts and sociological analysis together in a coherent and digestible account. The book contains a short "Lexicon of Archaic Terms," which features a fictional entry on "synthetic-failure paleoanalysis." This, we read, is a fictional scientific discipline that aims to understand "past failure, specifically by understanding the interactions (or synthesis) of social, physical, and biological systems" (62). *The Collapse*'s uses of narrative allows it to combine specific detail and general assessment, rigorous analysis and bold synthesis. Through its narrative form, in other words, *The Collapse* itself counts as a work of "synthetic-failure paleoanalysis."

Anthropocene Affect

The Collapse's shift to a post-civilization future not only produces analytical insights, it also aims to generate a particular emotional experience in its readership. The device presents contemporary civilization as something that can be lost, and thus instills a sense of vulnerability and contingency. The affective power of literature is a notoriously speculative issue, as particular formal choices are bound to affect different readers in distinctive ways. And how readers will translate affect into action is an even murkier issue: will an awareness of the fragility of civilization inspire actions that fortify that civilization, possibly at the expense of others? Will it inform the quest for previously unimagined forms of life? Or will it instill a sense of despair over human powerlessness in the face of overwhelming forces? As I explore in the second chapter ("Genres, Media, Worlds"), cognitive approaches to literature have tried to make our understanding of the emotional and social efficacy of literature more robust by grounding it in empirical observations. Starting from the insight that cognition is always embodied and affective, scholars have begun to explore how narratives about planetary devastation like Cormac McCarthy's *The Road* "immerse us into their storyworlds and engage us in the gruesome tale they tell about environmental disaster and human suffering" (Weik von Mossner 2017, 3). Analyzing how our bodies act as "sounding boards" for "characters' perceptions, emotions, and actions within … virtual worlds" (3), such research shows how literature's (but also film's and video games') "emotionalizing strategies" afford an immersive experience of ecological entanglements that more rigorously scientific writing cannot elicit (190).

Immersion is not the only way literature does its affective work: it also has the capacity to attune us to emergent realities that have not yet been consolidated as concepts and categories. The Anthropocene, we have seen, is less an accurate proper name than a necessarily provisional label for the ongoing effort to find new names, images, and stories to make sense of the bewildering changes to our sense of the relation between human and nonhuman lives. Before they are properly understood, environmental disturbances and social crises first register as sensory discomfort and affective unease. It is these disorienting feelings that literature and the arts can begin to tap into and record well before future historians will be able to codify our present with the benefit of hindsight. It is especially in the field of affect theory that what we can call literature's *seismographic* function has been elaborated. For a theorist like Lauren Berlant (2008a), literature can transmit affects that embed our bodies in the historical present; it can play its role in helping us experience "the present as an ongoing process and project of collective sensory detection" (2). For Berlant, such affective disruption becomes especially relevant when time-tested genres have ever less traction on emergent new realities. Recall how VanderMeer's *Annihilation* cultivates an unsettled mood that takes its bearing somewhere between science fiction, gothic, fantasy, horror, and the "old" weird, without conforming to any of these labels. *Annihilation*'s affective work does not operate within one particular generic template that programs the reader's emotive response, but decomposes and reconstitutes different generic elements to yield an intractable feeling that we can't really name, only experience. For Roger Luckhurst (2017) such "slid[ing] in and out of generic conventions" is a key feature of the weird. Its "waywardness" leaves readers confounded, and forces them to tap into moods and sensations through which historical changes register before they can be confidently categorized (1050). Literary experimentation, then, can attune audiences to modalities of the present that human reason cannot (yet) capture.

Anthropocene Imagination

If the weird or the stuplime, on which I touched in the first section, do not sound like very positive affects, they are not purely dysphoric either. Unlike the death-obsessed gothic mode of the uncanny, the weird, for Luckhurst, has less to do with the fear of death than with a dread of life—of the excessive, lush vibrancy of Area X, for instance. Dread is always "anticipatory, oriented toward the future," and signifies an openness to radically strange realities (Luckhurst 2017, 152). In the case of VanderMeer's trilogy, mesmerized fascination with a foreign territory gradually morphs into a curious symbiotic fantasy of cross-species intimacy. Existing discussions of literature and the Anthropocene tend to underline the *inadequacy* of existing literary forms for representing realities that are too gigantic, too slow, too dispersed, or too intangible to be fit into customary literary formats (Bond et al. 2017). It is also useful to stress literature's "world-making" capacities, its power to shape fictional realities that add unanticipated possibilities to the world in which these fictions emerge (Cheah 2016). We can see this at work in *Annihilation*'s many extensive descriptions of the environment. When the novel describes an underground tower (which is paradoxical enough in itself!) as "*breathing*," carrying "the echo of a heartbeat," and having walls "not made of stone but of *living tissue*" off of which "silvery-white phosphorence" is rising (41), it invites readers to imagine the border between living organism and dead matter differently and bring it in tune with a dispensation in which that boundary is not all that clear any more.

As I explore in my sixth and last chapter ("Residues"), Anthropocene literature's imaginative and world-making resources have been mobilized most intensively in stories set in the (near) future. William Gibson's 2014 *The Peripheral* is a science fiction novel whose world-building is technological, economic, and social rather than environmental (as it is in the *Southern Reach* trilogy). *The Peripheral* evokes a double future: one situated in the 2030s, in which the USA has deteriorated to the status of an impoverished trailer park populated by drug "builders," gamers, and cyborg veterans, and one situated in the twenty-second century, in which a small kleptocratic elite live hi-tech lives in gated communities for which they have the inhabitants of the "earlier" future provide cheap labor. This double future is an imaginative way to render the rampant inequality that besets contemporary society. The two futures are separated by an event called "the Jackpot": a massive civilizational collapse and planetary extinction event that brings together key Anthropocene anxieties. *The Peripheral* combines what Gibson calls "a more fully corrupt, third worlded version of contemporary America" (Newitz 2014) and a radically altered post-Anthropocene future. Margaret Atwood's wildly popular *MaddAddam* trilogy (consisting of *Oryx and Crake* [2003], *The Year of the Flood* [2009], and *MaddAddam* [2013]) also features two futures: one is marked by intensified consumerism and genetic experimentation, the other, following a global pandemic, offers new cross-species potentialities. This set-up also recalls the organizing conceit of *The Collapse of Western Civilization*, but what sets Gibson's (like Atwood's) work apart from Oreskes and Conway's more austere narrative exercise is his novel's indulgence in imaginative details: we find extensive descriptions of moving tattoos, of hi-tech suits and weapons, of a particular kind of time-travel through so-called "peripherals" (cyborg avatars users can connect to from a distance), and of bizarre cyborg wheelchairs. *The Peripheral* is a richly textured novel, and this shows its interest in an imaginative exploration of the lived reality of a possible future. This is something that Oreskes and Conway are simply not interested in: their focus is on a diagnosis of the present. In Atwood's trilogy, a lot of imaginative energy is invested in depicting the life of the Crakers, a bioengineered species of pseudo-humans. For all their strangeness and excess, these benign and naive creatures engaging in carefree group sex with their bright blue buttocks and erections add a weirdly (and ambivalently) utopian dimension to the books' critical engagement with present-day society.

Anthropocene Writing

One peculiar feature of the unruly entity that haunts *Annihilation*'s Area X is that it *writes*: "An ... organism ... was writing living words along the interior walls of the tower ... Whole ecosystems had been born and now flourished among the words" (VanderMeer 2014a, 90). The tower is filled with "a moldering pile" of documents chronicling earlier explorations of the terrain, which are gradually absorbed by the midden: "Torn pages, crushed pages, journal covers warped and damp. Slowly the history of exploring Area X could be said to be turning into Area X" (111–12). The first question this raises is one of *agency*: what, the novel asks, "was the interplay between the words and the tower-creature ... Were the words a form of symbiotic or parasitic communication between the Crawler and the Tower" (91)? The theme of writing brings into relief the agency of nonhuman actors; if we define agency as the capacity to have an impact, to leave traces for others to read, then it makes sense to figure agency as, precisely, a form of writing. And as literature is a kind of textuality that is deeply concerned with the powers, the limits, and the (im)possibilities of writing, this offers a fourth way—after narrative, affect, and imagination—in which a consideration of literature can enrich our understanding of the Anthropocene. Without going so far as to say that the Anthropocene is essentially a *literary* problem, we can submit that literature helps us see to what extent the Anthropocene is a matter of reading and writing, of decoding and inscription.

In the last decades of the twentieth century, Jacques Derrida's (1974) infamous assertion that "il n'y a pas de hors-texte"—best translated as "there is no outside of the text" (158)—used to function as a slogan for a form of cultural studies that believed that the whole of reality could be studied as a text. It was easily ridiculed as an example of so-called High Theory's (or French Theory's) stubborn blindness to material reality. In a time defined by the human capacity to promiscuously leave traces in the chemical and climatological make-up of the planet and by nonhuman agents' ability to also leave their mark, Derrida's statement today seems perversely appropriate. There is no part of the earth system that is *not* affected by the traces our daily actions leave in the atmosphere; nor is there any aspect of our surroundings that we, together with our technological devices, are *not* frantically scrutinizing to harvest data about the world. Powered by our ever increasing technological capacity to read geological, chemical, and climatological traces, the Anthropocene world is a reality saturated by the almost boundless proliferation of data, inscriptions, and signs (what has been referred to as "infowhelm" or, more accurately, "semiowhelm" [Woods 2017, 205–206]); it is a world that is manically read even as our actions provide ever more data points for future readers.

Writing is not just a reflection *on* the Anthropocene, it is also *constitutive of* it. The Anthropocene, Tobias Boes and Kate Marshall (2014) remark, "is not simply something that is written *about*; it is also something that is actively shaped and created through acts of human inscription" (64). Human actions, in other words, do not function as denotative speech acts but are "performative interventions in which humankind functions as both subject and object" (64): the actions we undertake as *subjects*, together with the actions of nonhuman agents, make up the reality of a geological epoch that carries our name, and that thus defines what we (as *objects* of these speech acts) are. This is a difficult idea, to be sure. Self-reflexive experiments that test the paradoxes and limits of writing are a hallmark of modern literature, and especially of the high modernist projects of writers like Flaubert, Mallarmé, Joyce, Beckett, and Blanchot. Literature's capacity to interrogate its own conditions of (im)possibility gains a surprising new relevance in an era in which there is nothing that is not marked by both human and nonhuman traces.

In a digital age, the media through which we record reality are as much a part of our lived environment as the natural world. In the second volume of the *Southern Reach* trilogy, entitled *Authority* (2014), the setting shifts from the weird wilderness to the office. Control, the novel's protagonist, is charged with investigating the Southern Reach's dismal track record. Rather than a riveting procedural, however, *Authority* is a comedy of data management, in which Control is overwhelmed by the proliferation of writing: "Even when he asked questions he was hemorrhaging data. He had a sudden image of information floating out the side of his head in a pixelated blood-red mist" (VanderMeer 2014b, 142). Data bleed into the world and further erode the distinction between human and nonhuman realms. The writings he has to put in order "looked in part as if he had tracked in dirt on his shoes from outside," turning him into "a new kind of urban farmer, building compost piles with classified material" (152). Gradually, the difference between the settings of the first and second volumes collapses as the sprawling mess of information becomes an environment in its own right: "His office began to close in on him. Listless pushing around of files and pretend efforts to straighten bookshelves had given way to further Internet searches" that provide images that look like Area X (288). It seems that digital media and natural environments are deeply entangled in our experience of the world.

Developments in media technology are a crucial dimension of the Anthropocene. Registering changes to our lifeworld requires increased capacities to read, for which we mostly depend on technological and computational aides. Changes to the earth system such as rising temperatures and carbon dioxide levels can only be observed through "computational models supported by a knowledge infrastructure" such as "weather observations, satellite data, radar readings, and so on" (Mirzoeff 2015, 219). A more perverse link between digital technology and environmental degradation is the fact that these technologies leave a considerable environmental footprint through their vast expenditure of (often unclean) energy as well as through the use of rare minerals and the proliferation of e-waste (Parikka 2015, 111–13). Making planetary change visible, in other words, depends on computational processes that have a material impact which, in a vicious feedback loop, contributes to that change.

The entanglement of digital and natural ecologies means that literary works that are not explicitly interested in environmental issues can still tell us something about the Anthropocene. In Tom McCarthy's novel *Satin Island* (2015), U., the novel's narrator, works as a corporate anthropologist charged with writing the "Great Report": an all-encompassing, comprehensive account of contemporary life. Given this daunting task, it is unsurprising he hits a wall: "I'd begun to suspect," he notes, "that this Great Report was un-plottable, un-frameable, un-realizable: in short, … *un-writable*" (McCarthy 2015b, 126). U. gains insight into his impasse when he begins to understand that, in a data-saturated world in which physical movements, consumer transactions, keystrokes, and click-throughs are relentlessly recorded, tabulated, and cross-indexed, the Great Report is not so much "un-writable" as *being written in real time*:

> The truly terrifying thought wasn't that the Great Report might be un-writable, but—quite the opposite—that it had already been written … by a neutral and indifferent binary system that had given rise to itself, moved by itself and would perpetuate itself … And that we, far from being its authors … were no more than actions and commands within its key-chains.
>
> (133–34)

In a world pervaded by writing, *human* agency is no longer exceptional, as it is overwhelmed by nonhuman and technological forms of writing. In an essay, McCarthy (2015a) notes that the algorithms of Google and Facebook have taken on many of the tasks traditionally allotted to literature: they transcribe human action into a "regime of signals" that is "omnipresent and insistent … undeniably inserted or installed at every stratum of existence."

Even if *Satin Island* is not primarily interested in the natural environment, its sustained focus on processes of writing turns it into an exploration of the entanglement of human and nonhuman forms of agency. The novel reminds us that processes of notation and registration, even if automated, are constitutive of contemporary life: in Mark Seltzer's (2016) words, the contemporary world consists "both of itself and its self-description, denotation, or registration" (6). This points to the fact that, in a digital age, everything that happens is obsessively and instantaneously archived. Following Derrida (1995), this compulsion is often referred to as "archive fever": a manic recording of the present in the face of its imminent disappearance. The Anthropocene has, if anything, exacerbated our sense of the fragility and impermanence of things. But Seltzer's words and McCarthy's novel make the further point that archiving is not just a neutral act of registration, but an intervention that actively shapes reality. Google's algorithms' knowledge of behavioral patterns allows it to predict and thus to influence and capitalize on our consumer preferences. In the context of the Anthropocene, we can see that this dynamic is not very different from the way in which the atmosphere's archiving of human carbon dioxide consumption aggravates the heating of the planet. In the words of Mark Seltzer again: "It is not merely that there is nothing in the world that is not in the files: the correlate is that there is then nothing in the files that is not in the world" (144). All the world's a file, and all the men and women merely agents composing that file: at the juncture of literary and environmental thinking, the complex dynamics of writing are as important as the powers of storytelling, affect, and the imagination.

The Book: Anthropocene Agencies, Anthropocene Temporalities

The Anthropocene has already proven to be a fertile new context for the study of literature. Not only is there an intense dialogue between literary criticism and contemporary literature that engages with environmental issues, but the Anthropocene has already made it possible to revisit the literary historical archive and mine texts we thought we had exhausted for intimations of nonhuman agency, geological timescales, and cosmic disturbances. But the traffic between literature and the Anthropocene is emphatically not a one-way street: literature's affective, narrative, imaginative, and self-reflexive powers make it a vital site for coming to terms with the challenges of the current environmental crisis. As I elaborate in my first chapter, this makes literature an important participant in the project of the environmental humanities, a capacious field in which environmental thinking across disciplinary borders is currently thriving. In critically mapping the interface of literature and the Anthropocene, this book addresses two overlapping constituencies: students and scholars of literature, who can profit from the coordinates and trajectories this book provides to help them navigate the exciting new intellectual terrain the Anthropocene makes available; and students and scholars working in the environmental humanities, who may discover that literature and literary studies offer unexpected resources to enrich ongoing debates on the planetary crisis.

This introduction has already argued that the Anthropocene extends the range of agents that compose the world; it has also pointed to its drastic reorganization of our experience of time, as human life encounters vast geological time scales and the open future we took for granted threatens to be foreclosed in the face of climate disaster. The rest of this book is organized around these two changes: the first part ("Anthropocene Agencies") presents the new agents comprising the Anthropocene world, while the second part ("Anthropocene Temporalities") focuses on the way that world reorders the relations between past, present, and future. Cumulatively, the two parts survey key concepts and ideas as well as literary examples in ways that extend our vocabularies for confronting environmental degradation. It often seems that Anthropocene discourses about the nonhuman and the posthuman, about hyperobjects and vibrant matter, about apocalypse and catastrophe (to name but a handful of key terms) are disconcertingly interchangeable and frustratingly self-referential. This book will draw on literary examples to insist on the specificity of such terms and demonstrate their analytical potential.

Chapter 1 ("Forms, Lives, Forms of Life") situates discussions of literature and the Anthropocene in their institutional context. Drawing on the example of a contemporary form of nature poetry called ecopoetics, and especially the work of poet Evelyn Reilly, I trace how the literary critical domain of ecocriticism has increasingly come to join forces with other disciplines under the rubric of the environmental humanities. I propose that the notion of *form*, which has a close affinity to literature, can help us orient ourselves among the proliferating agencies that overwhelm traditional ways of apprehending the world. I introduce a distinction between (biological) life forms and (cultural) forms of life to show how literature's *life of form* can help make sense of the bewildering multiplicity of forces that afflict us, and maintain distinctions between different differences: between cultures, but also between different biological life forms.

Chapter 2 ("Genres, Media, Worlds") homes in on the literary critical notion of *genre*. Genres are not only templates for organizing our experiences of the world, they are also affective scenarios through which we orient ourselves in a world whose coordinates we haven't yet fully understood. Recent literary scholarship has mostly been critical of the capacities of literary realism to capture the unruly realities of the Anthropocene, and has tended to affirm the relevance of historically disparaged genres such as science fiction and horror. I survey approaches that aim to map particular formal features onto specific affective reader responses, before I turn to a comparative case study on literature about the Arctic, with a focus on Edgar Allan Poe's *Arthur Gordon Pym*, to argue that it might be the friction between generic templates that provides the best intimation of a nonhuman reality. The rest of the chapter links this intimation to the conceptual differences between worlds, globes, and planets, and explores the affordances of different media to cultivate such apprehensions of a nonhuman dimension to life. All these media make us apprehend an alien reality, and they do so in complementary and media-specific ways.

In Chapter 3 ("Objects, Matters, Things"), I critically introduce a number of influential theoretical paradigms for thinking the nonhuman world. Exploring the distinctions between *object-oriented ontology* and the *new materialism*, I turn to the example of John Burnside's novel *Glister* to assess the relative affordances of these approaches, and to show how they ultimately fall short of the task I elaborate in chapter 1: that of differentiating different differences. After linking this challenge to the all-important issue of scale, I conclude with a discussion of Serenella Iovino and Serpil Oppermann's *material ecocriticism*. Where these thinkers articulate language with the material world through the notion of "storied matter," I reiterate the point I adumbrated in the previous section: even more than storytelling, it is writing that offers a literary model for thinking about the Anthropocene world.

With these diverse agents in place, the book's second part turns to the different temporal vectors that make up our experience of the unstable Anthropocene present. In order to find a new vocabulary to account for the intimations of futurelessness, of radically expanded timescales, and of a strange mixture of urgency and stasis, I take my inspiration from the work of cultural critic Raymond Williams. In his *Marxism and Literature*, Williams (1977) influentially argues that our experience of the present is never exhausted by dominant elements alone: the present always also contains what he calls "residual" elements that have not been fully incorporated in the dominant order as well as "emergent" forces that have yet to receive a consolidated form (121–27). The critical task, as Williams sees it, is to capture "the dynamic interrelations, at every point in the process, of historically varied and variable elements" (121).

Williams accords an important role to art and literature in confronting "the undeniable experience of the present … the specificity of present being" (128). This specificity manifests itself as a "structure of feeling" rather than a full-fledged ideology or worldview. Literature and art inhabit "[a]ll the known complexities, the experienced tensions, shifts, and uncertainties, the intricate forms of unevenness and confusion" (129), and can provide access to "social experiences *in solution*" (133). Williams's account of the complex and dynamic nature of the present and of the coincidence of nonsynchronous forces resonates with the Anthropocene's challenge to our customary modes of making sense of life. At the same time, the linearity suggested by the triad residual/dominant/emergent suggests a sense of orientation, even progress, that is decidedly out-of-sync with the more radically disjointed present of our planetary crisis.

The three terms I use to update Williams' triad and that collectively capture the Anthropocene's disjointed temporality are dominations, emergencies, and residues. Chapter 4 ("Dominations") develops the notion of Anthropocene *memory* to underline the persistence of a long and slow history of environmental violence. Drawing on a revised version of Walter Benjamin's famous angel of history, I explore how the context of the Anthropocene affects the objects, the subjects, as well as the archives of memory. I read two contemporary novels, Thomas King's *The Back of the Turtle* (from Canada) and Alexis Wright's *The Swan Book* (from Australia) that draw on Indigenous archives to elaborate an environmental memory. This memory is not only oriented to the past, but also to a foreclosed future. I present notions such as solastalgia, ecological grief, and pretrauma to show how the study of memory contributes to a vocabulary for present and future disappointments and discontents.

Another temporal dimension of the Anthropocene present is the disabling combination of a sense of urgency and inaction. Chapter 5 ("Emergencies") explores this paradox through a focus on *infrastructure*. Infrastructure names the utilities—roads, electric grids, sewage systems—that allow human life to thrive and that are simply supposed to do their job. Infrastructure only calls for attention when it fails. Drawing on a number of literary examples—Jamaica Kincaid's *A Small Place*, Karen Tei Yamashita's *Tropic of Orange*, and Ben Lerner's *10:04*—I show how the temporality of infrastructure is more complicated than the dyad of invisibility and emergency suggests. By foregrounding the crucial importance of energy infrastructures, I draw on insights from the interdisciplinary field of the energy humanities to show that infrastructure can even become a catalyst of hope and provide a sense of continuity and a shared experience that can help us confront climate emergency without resorting to apocalyptical thinking.

Finally, in Chapter 6 ("Residues"), I focus on a final element in the complex Anthropocene structure of feeling: a sense of futurelessness powered by a fear of imminent species extinction. I show how literature and literary theory have not only been concerned with this theme, but also with the (slightly different) fear that the human will survive its own demise as a mere residue, as a trace whose meaning it can no longer control. Shifting the issue from a fear of extinction to an anxiety over human insignificance, I survey four affective dispositions through which contemporary literature and theory confront that anxiety: denial (which I observe in the wildly popular genre of the post-catastrophe novel), detachment, indifference, and outright misanthropy. I find scattered traces of hope in literary works by Ben Lerner, Donna Haraway, and Richard McGuire that manage to achieve a certain disengagement from forms of life that are fast becoming unsustainable. If only because it contains such traces, literature itself is a form that does not deserve to become obsolete just yet.

Part 1

Anthropocene Agencies

1 Forms, Lives, Forms of Life

Ecopoetics and Form at the End of Nature

One important implication of the Anthropocene is that nature, as a reality that is divorced from human culture, no longer exists. Humans' far-reaching impact on the planet means that no part of the environment is untouched by human action, and that the world is now a thoroughly contaminated and compromised place. It would be wrong to see this is a sudden change: one of the lessons of the Anthropocene is that human and nonhuman realities have *always* been entangled. For many premodern communities, the idea that nature and culture were *not* intimately interlinked would not have made sense. The opposition between natural and cultural realms was always a cultural construct, and part of a process in which literature played a crucial role. Inherited ideas about nature were partly shaped by literary forms: the genre of the *pastoral* constructed the countryside as a repose from urban environments; in *elegies*, the cyclical rhythms of nature compensated for human losses and deaths; *westerns*, for their part, delivered the environment as a wilderness to be conquered and subdued. In the face of the Anthropocene, literature is revising these traditional forms and even developing radically new ones. In this chapter, I explore how formal innovation aligns contemporary literature with the aims of the interdisciplinary field of the environmental humanities.

One of the most remarkable works of Anthropocene poetry, Evelyn Reilly's 2009 collection *Styrofoam*, opens with the lines "Answer: Styrofoam deathlessness / Question: How long does it take?" (9). What outlasts human impermanence here is not the soothing rhythms of natural regeneration but the "deathlessness" of thermoplastics. Synthetic polymers like polystyrene (of which Styrofoam is a brand name) are gradually replacing animal and plant life in the world's oceans, especially the Atlantic and the North Pacific, and they are there to stay: nonbiodegradable, they reach beyond biological into geological time and will remain part of the landscape that nature poets of the (far) future will have to reckon with. In Reilly's poem, "Styrofoam deathlessness" is a nonnegotiable fact—it is the answer that precedes whatever questions we want to direct at it. The second page of the poem restates this fact: "Answer: It is a misconception that materials / biodegrade in a meaningful timeframe // Answer: Thought to be composters landfills / are actually vast mummifiers // of waste // and waste's companions // lo stunning all-color // heap-like & manifold.of // **foam**" (10). These two answers are neither preceded nor (as on the first page) followed by a question. In a sense, adding a question to what is an unalterable fact would be a way to pretend that such a reality somehow answers to human concerns. The poem's elision of the question underlines that such human considerations are beside the point, and that Anthropocene poetry needs new ways to reflect a denatured world. The distorted order and punctuation of Reilly's poem ("heap-like & manifold.of") similarly points to the need to find a new syntax and new forms for an altered reality.

Plastic in many ways condenses the paradoxes of the Anthropocene. Unsurprisingly, the alternative label of the "Plasticene" has already been coined (Reed 2015). Cheap to produce and available for many uses, plastic became the key substrate for consumer capitalism as it developed after the Second World War. Heather Davis (2015) explains that plastic embodied "the promise of sealed, perfected, clean, smooth abundance" (349). Plastic's shiny surfaces fostered the fantasy that we could disconnect from the recalcitrant, amorphous, and inconvenient demands of the natural environment and seal ourselves off in a self-contained bubble (354). Yet plastic has refused to serve as a clean barrier that shields us from contingency, as its own proliferating material reality has come to destroy ocean and animal life and expose our human bodies as permeable and penetrable. In Reilly's collection, plastic infiltrates human bodies: "Enter: 8,9,13,14,17-ethynyl-13-methyl- / 7,8,9,11,12,14,15,16-octahydro-cyclopenta-diol // (aka environmental sources of hormonal activity / (side effects include tenderness, dizziness / and aberrations of the vision" (9). The brackets never close, nor do the brackets within the brackets, as if to underscore that the human body, the female sex hormone estrogen (encoded in the chemical formula), and thermoplastics (which infamously release estrogenic chemicals) are inseparably enmeshed in an open circuit. If "tenderness" and "vision" are traditional attributes of poetic language, in this new constellation they are merely the side effects of plastic-induced hormonal disturbances. Anthropocene environments are not just sublime spectacles to be contemplated (there is a strange beauty to floating Styrofoam islands), they also course through human bodies. Human corporeality is, in theorist Stacy Alaimo's (2010) words, always a "trans-corporeality," in which "the human is always intermeshed with the more-than-human world ... 'nature' is always as close as one's own skin—perhaps even closer" (2).

Styrofoam draws inspiration from a long tradition of nature writing, even if it sees traditional forms as deeply implicated in the destruction of the environment. Reilly's opening poem references Samuel Coleridge's famous "Rime of the Ancient Mariner," a tortured reflection on a sailor's gratuitous killing of an albatross and a keystone of British romantic poetry. Coleridge's "For all averred, I had killed the bird" becomes "*(for all averred, we had killed the bird* [enter albatross / stand in of choice" (11). Responsibility, in Reilly's poem, is shared by a collective "we," and the albatross is eerily interchangeable with other targets of human overreach—not least the seabirds choking on plastics that have become powerful emblems of the Anthropocene. Coleridge's "Rime" dissects how the fallout of that transgression comes to haunt the sailor for the rest of his life; it shows how human life in overextending itself also ends up targeting itself. Reilly's collection also amply references Wordsworth, another central figure in the consolidation of the romantic celebration of nature, and Herman Melville, whose *Moby-Dick* precociously intimated the dangers of planetary profiteering (exemplified by the whaling industry), while it also perpetuated romantic associations between nature and transcendence. As Lynn Keller (2015) has noted, *Styrofoam* shows how the romantic habit of finding consolation in a sacralized nature, by denying the entanglement of human and nonhuman entities, is complicit in the degradation of nature to which the collection testifies (854–56).

"*(for all averred, we had killed the bird* [enter albatross": Reilly mobilizes italicization and (square) brackets to navigate a sea of information, a surfeit of seemingly nonbiodegradable signs. The environment in which contemporary poetry operates also contains digital flotsam: "*Monica T / Soft and satisfying for infant teething if you first freeze.* / posted 10/11/2007 at thriftyfun.com / ... All this. information / anddeformation // & barely able to see sea" (11). The Internet is not only a platform that stimulates the consumption and production of ever more thermoplastics that stubbornly refuse to disappear, it also generates an "infowhelm" in which poetry threatens to drown. The intrusion of so many discourses—from chemistry over cultural history to the everyday banality of customer reviews—in the body of the poem is crucial for *Styrofoam*: it is not only a collection about environments after the end of nature, but also about the viability of poetry in saturated information ecologies.

In *Styrofoam*, the consolations of pastoral make way for formal innovations that signal a crisis that is environmental as well as literary. Reilly's work is often considered as a work of *ecopoetics*, a kind of writing that directly addresses environmental crisis and that emerged in the 1990s and was later consolidated around poet and critic Jonathan Skinner's experimental journal *ecopoetics*. For Juliana Spahr (2011), another poet affiliated with this development, ecopoetics differs from traditional nature poetry that, as she writes, "even when it got the birds and the plants and the animals right it tended to show the beautiful bird but not so often the bulldozer off to the side that was destroying the bird's habitat" (69). In ecopoetics, environmental crisis is also a crisis of representation, as the customary formats that organized the divisions between the human and the nonhuman are unable to accommodate both bird and bulldozer. Ecopoetical form, in contrast, reflects what Margaret Ronda (2014) calls "incommensurabilities and violent estrangements" (105). In the case of *Styrofoam*, the inevitable inadequacies and distortions of representation are reflected in the unstable lay-out, the erratic punctuation, the quasi-misprints, and the agglutination of words ("All this.information / anddeformation"). The collection resembles nothing so much as the effect of copy-and-pasting an image-based PDF as plain, editable text. The collection's most remarkable stylistic features are then so many glitches afflicting this attempted conversion.

There are two ways to assess *Styrofoam*'s formal innovations: as evidence of a breakdown of obsolete forms, or as an attempt to design new forms for a new reality. Most existing literary criticism on the Anthropocene tends to underline formal failure, and to promote works whose "material and formal qualities … come to displace and overwhelm" established forms of expression and representation (Clark 2015, 183). It emphasizes "disjunctiveness, a being-overwhelmed by contexts in which the human perceiver is deeply implicated but cannot hope to command or sometimes even to comprehend" (184). This emphasis on formal inadequacy in the face of crushing realities recalls the structure of the romantic sublime: as I already explored in the introduction, this model stages a collapse of the human capacity to represent an overpowering reality (mountains, waterfalls, canyons) but still snatches some consolation from its higher-order insight in this mismatch between mind and matter. It is certainly possible to describe Reilly's glitchy collage in those terms. At the same time, we also saw her poetry dismiss such a recuperative sublime in its critical update of Coleridge, Wordsworth, and Melville.

It may be more productive to consider *Styrofoam* as an ultimately successful effort to devise new forms rather than as a campaign to break obsolete ones. The collection's capacious collage of human and nonhuman forces, of different commercial, scientific, and cultural discourses, but also of images, charts, drawings, and diagrams *is* a formal achievement: it presents a formal equivalent of the sprawling, weirdly unexciting, but increasingly unavoidable presence of thermoplastics in the planetary ecosystem. Old forms (such as the romantic sublime) are not displaced or discarded: they are recycled as part of this new formal construction. The text's sparse and irregular lay-out of scratches and dots on a white surface even resembles the typical texture of Styrofoam. This is not the naïve evocation of a text-external reality typical of traditional poetry, but a form of mimesis that locates itself in the Anthropocene infowhelm and mobilizes the affordances of poetry to convey, in different ways, the contours and the feel of a previously unsung but increasingly inescapable environmental reality. *Styrofoam*'s structural, visual, typographic, and stylistic choices all point to an attempt to make rather than break form.

The notion of form can help us understand how literature can contribute to the interdisciplinary study of the Anthropocene under the label of the environmental humanities (a field I map in the third section of this chapter). The Anthropocene is marked by an intense reorganization of the relations between different (cultural, political, social) *forms of life* and (biological) *life forms*—an unstable distinction to which I return below. A medium like literature that is defined by its restless grappling with form can help us gain traction on this mutating reality. Of course, form is a notoriously slippery notion. Literary criticism can be relied on to rediscover the importance of form every few decades. Since the 1990s, the discipline has launched several "new formalisms" (Levinson 2007), which has not resulted in a shared understanding of the term (Kramnick and Nersessian 2017). Yet this lack of a strict definition also makes form a versatile and flexible term for articulating complex realities. In her book *Forms: Whole, Rhythm, Hierarchy, Network*, Caroline Levine (2015) emphasizes that the notion of form not only points to artistic and literary arrangements, but also to social ones. For her, "form always indicates *an arrangement of elements—an ordering; patterning, or shaping*"; form, for Levine, means "all shapes and configurations, all ordering principles, all patterns of repetition and difference" (3). Forms "can organize both social and literary objects" (13), and a sustained focus on form makes it possible to trace the ways literature engages social life through the collisions, overlaps, and intersections of forms.

Anthropocene literature does not stop at social life. As it also engages biological, geological, and cosmic dimensions, it traffics in things that are precisely *not* configurations, that are *not* ordered, but rather intractable, resistant, diffuse, or even intangible. Tracing encounters with such things requires a more capacious account of the relations between form and life. In a felicitous phrase, *Styrofoam* calls the Anthropocene "our infinite plasticity prosperity plenitude" (Reilly 2009, 43), suggesting that in the Plasticene, the notion of life itself has become plastic. Following the work of Catherine Malabou, Arne De Boever (2016) explains that plasticity does not mean that life is infinitely malleable and fungible, but that it is able to "receive, give, and explode form" (24). But Anthropocene form does not stop at life: following Elizabeth Povinelli's (2016) work on "geontopower," we can see that it is not only life that is at stake, but, "as the previously stable ordering divisions of Life and Nonlife shake," we also need to be mindful of the differences between *bios* and *geos*, between living and nonliving things (5). Some things are precisely *not* malleable, not plastic. One of these things is, surprisingly, plastic itself. Heather Davis (2015) remarks that plastic may be "the hardest material there is," as "it refuses its environment, creating a sealant or barrier that remains impermeable to what surrounds it" (351–52). Even when plastic breaks apart, "[t]he molecules themselves remain intact, holding onto their identity" (352). A sufficiently capacious account of form, in other words, also needs to factor in a resistance to formal patterns and arrangements.

In the next section, I situate the relation between life and form in the development of the field of ecocriticism, the literary subfield in which that relation has been discussed since the early 1990s. As I show, ecocriticism traditionally celebrated literary writing for its capacity for literary mimesis, before it took inspiration from a strain of critical theory it initially resisted to move on to a more sober appreciation of the constructedness of nature. Only more recently has ecocriticism shifted from the extremes of mimesis and construction to a consideration of the multifarious interactions, not only between literature and science, but also between the natural and the social world. In the section after that, I draw on the distinction between forms of life and life forms to show how contemporary ecocriticism, this time increasingly operating under the banner of the environmental humanities, negotiates literature's engagements with *different differences*: with different histories, values, ethnicities, and classes, but also with natural and geological forces, and with kinds of life and nonlife. Giving these different elements their due involves remaining attentive to their specificity and not collapsing them into all too vast categories such as "the other" or "the nonhuman."

To illustrate the importance of differentiating differences, the last section of the chapter turns to the field of critical ocean studies with a discussion of J.M. Ledgard's novel *Submergence* and his curious *Terra Firma Triptych: When Robots Fly*. These works exemplify the interactions between biological and social forms and demonstrate the complementary needs for scientific literacy and cultural knowledge. Both works are thematically complex, draw on different forms of disciplinary expertise, and deploy multiple genres and formats. They illustrate how the Anthropocene challenges the limits of ecocriticism and brings the ecological study of literature into the remit of what Ursula Heise (2014) has called "the transdisciplinary matrix" of the environmental humanities. Ecocriticism's contiguity to disciplines such as environmental history, environmental anthropology, cultural geography, critical animal studies, and political ecology (all of which increasingly operate under the umbrella of the environmental humanities) becomes a vital asset when confronting the Anthropocene.

What Was Ecocriticism?

One curious fact about the history of academic ecocriticism is that it took so long to catch up with non-academic environmental writing. *Silent Spring*, Rachel Carson's 1962 condemnation of pesticide overuse, is commonly considered the beginning of modern environmentalism (Garrard 2004, 1), and the two-page "Fable for Tomorrow" that opens Carson's text already prefigures Anthropocene discourses in uncanny detail. There is a (decidedly suburban) pastoral (the book begins: "There once was a town in the heart of America where all life seemed to live in harmony with its surroundings"), a sudden apocalyptic interruption ("Then a strange blight crept over the area and everything began to change"), echoes of radioactive fallout (as "a white granular power" falls "like snow upon the roofs and the lawns, the fields and the streams"), and finally the realization that responsibility cannot be outsourced to "witchcraft" or "enemy action": "The people had done it themselves" (Carson 2002, 1–3). Carson's opening fable is marked by an astute use of formats such as pastoral and biblical apocalypse and by the multidirectional interactions between technology, nature, and human life. This shows that the Anthropocene is only one chapter in a longer history of ecological awareness, making it all the more remarkable that such awareness initially found few echoes in the academic study of literature.

In the years between *Silent Spring* and the emergence of ecocriticism in the early 1990s, and definitely since the 1970s, literary criticism in the Anglophone world was dominated by a philosophically informed consideration of the ways texts construct realities and identities. Representation and narrative were studied as processes that shape the world; because such processes were indelibly warped by power, the capacity of literary texts to reliably refer to a text-external reality met with skepticism at best. The use of biological metaphors (as when the state is imagined as a body politic, or feminine beauty as a flower) was seen as a sinister form of "naturalization": invoking the stability of nature was seen as a ploy to pass off mutable social arrangements as self-evident and common sense. Until the 1950s, the romantic notion of literary works as organic and internally harmonious totalities had dominated literary criticism. The living harmonies of literature were categorically separated from what were seen as the reductive and soulless operations of science and theory. The critical movements that followed (especially poststructuralism, deconstruction, Marxist criticism, and the new historicism) reacted to this overly pious approach to literary objects and drew on theoretical and non-literary knowledge to dismantle rather than celebrate literary texts. In this transition, the natural world stayed out of the picture. To the extent that literary criticism engaged political issues (and it did so increasingly in the 1980s and 1990s), it attended to the recognition of ethnic, gender, and sexual minorities rather than to questions of environmental justice.

Only more recently has literary criticism begun to see that the constructedness of nature and nature's irreducible materiality are not mutually exclusive ideas, and neither are the value of nature and the importance of cultural difference. The development of ecocriticism is customarily figured as a series of (as many as four) "waves," an image that does not imply "a tidy, distinct succession" so much as continuities and reverberations that are hard to tell apart (Buell 2005, 17; Marland 2013; Slovic 2010). The insistence on the physical reality of nature, for instance, has persisted across all four waves. In the *first-wave ecocriticism* of the 1990s, the rescue of nature (from mistrustful new historicist and Marxist readings) generally went hand in hand with a reappraisal of nonfiction nature writing, which was credited with the power to vividly evoke the natural world: the lyrical poetry of a romantic like Wordsworth (in the UK) or the environmental imagination of native American traditions or writers like Henry David Thoreau and Wendell Berry (in the USA). The backlash against three decades of theorizing was reflected in a definite hostility to literary theory's alleged "pantextualism": its belief that only texts mattered, and that the real world only mattered to the extent that it could be read as a text. Early ecocriticism countered this theoretical and textual focus with what critic Lawrence Buell (1995) calls "a spirit of commitment to environmentalist praxis" (430). Nor was theory the only culprit: because of its tendency to objectify nature and its proximity to technology, urbanization, and industrialism, science became a prime target in ecocriticism's wholesale critique of modernity (Heise 1997, 4–5).

Ecocriticism's rediscovery of nature proved more enduring than what Buell (2011) calls its initial "naively pre-theoretical valorization of experiential contact with the natural world and its trust in the power of artifacts either to render the natural world or motivate return to it" (94). *Second-wave ecocriticism* overcomes the anti-theoretical and anti-scientific bias that characterized the field in its early days: it graduates from Thoreau and Wordsworth to a writer like Carson, from the untainted wilderness to compromised sites of industrial and urban transformation, and from a celebration of ecstatic immersion to a more sober recognition of the differential vulnerability of various constituencies. Ecocriticism even moves beyond Carson's suburb to postcolonial and Indigenous (and not only native American) territories, as it teams up with feminism and postcolonialism to pose questions of environmental justice and to foreground that women and other non-dominant groups are disproportionally affected by environmental deprivation (Marland 2013, 852). Scientific evidence is no longer the enemy and becomes a crucial ally in understanding the complexity of ecosystems and in claiming the body as a vital part of the affected environment.

In the new century, this broader literary canon, more inclusive understanding of the environment, and renewed appreciation of interdisciplinary knowledge has informed two further developments. A *third wave* of ecocriticism has amplified the resolutely transnational outlook that was already part of the second way. It has increasingly complemented the recognition of environmental commonalities with an appreciation of ethnic and national diversity (Adamson and Slovic 2009, 6–7). Coupling an ecological appreciation of transnational interconnectedness to an attention to cultural difference that is key to literary and cultural studies, ecocriticism has since the beginning of the century increasingly begun to pay attention to the ways Northern-hemispheric patterns of consumption impact the Global South (as the site of an "environmentalism of the poor" [Martinez-Alier 2002; Nixon 2011]). It has increasingly welcomed ethnic perspectives on natural deprivation—which is to say that it has again expanded the ecocritical canon—and advanced the notion of "translocality" to study the entanglements that generate the sense of place that first-wave ecocriticism simply took for granted. Third-wave ecocriticism has challenged earlier ecocritics' default reliance on bioregionalism, which insists on the reciprocal attunement between human cultures and their local surroundings, and the related idea that environmental commitments are fostered by an attachment to place. Instead, it has explored the possibilities of truly planetary imaginaries and an "eco-cosmopolitanism" that resituates "imagined communities" from the level of the region or the nation to larger scales more appropriate to the daunting challenges of the Anthropocene. Such a *planetary* ecocriticism operates within the tension between cultural difference and a shared species history without surrendering either of these commitments. What is needed, Ursula Heise (2008) writes, is "effective aesthetic templates by means of which to convey … a dual vision of the Earth as a whole and of the different earths that are shaped by varying cultural contexts" (210).

Third-wave ecocriticism has reminded the field of its earlier ethnic and cultural blinders and has allowed it to develop a more diversified landscape in a resolutely global framework. A f*ourth wave* has extended this appreciation of difference beyond the limits of the human. A properly inclusive account of the environment also features agents that are neither human nor natural; if ecocriticism emerged as the field in which "[t]he nonhuman environment is present not merely as a framing device but as a presence" (Buell 1995, 7), then it must also attend to unglamorous things like Styrofoam, dirt, stones, infrastructures, meat, microbes, and so on. Scott Slovic (2012) refers to this as "the fourth-wave material trend in ecocriticism" (619). The emphasis is no longer on humans' transformative power only, but on the distributed agency that brings together human, technological, natural, and other nonhuman actors. If third-wave ecocriticism envisioned forms of eco-cosmopolitan connectedness and what Lawrence Buell (2007) calls "ecoglobalist affects" (227), this strain imagines community beyond the boundaries of the species in multispecies assemblages.

In the early 1990s, ecocriticism emerged as an effort to attend to the interests of the nonhuman world, and to mobilize (an initially all too restricted canon of) literary writing to solicit such attention. A few decades on, it finds the human concerns it strategically bracketed and later reintroduced in the name of cultural difference fatefully enmeshed with nonhuman forces. The human has not so much been naturalized as become posthuman, just as the body has become, in Stacy Alaimo's (2014) words, "trans-corporeal," sustained in "the material interchanges across human bodies, animal bodies, and the wider material world" (187). As Pippa Marland (2013) remarks, this points to "a shared materiality between the human and non-human world that renders obsolete the distinctions between human and environment, moving beyond the construct of 'nature' altogether" (856). This decentering of human agency is often discussed under the rubric of the posthuman (a topic I engage in more detail in the third chapter). Some versions of posthumanism are decidedly cosier than others: on the one hand, there is the promise of an unsuspected intimacy with nonhuman animals, with viruses, bacteria, and microbes, and even with plants (studied in the field of critical plant studies); on the other, there is the more sobering realization that, in an age of biotechnological manipulation and digital connectedness, human life might be no more than a relay in media circuits and cycles of production and consumption, and that whatever agency we thought we had has been surrendered to machines or to thermoplastics upsetting our hormonal balance.

If first- and second-wave ecocriticism are defined by attitudes of celebration and critique respectively, the sheer complexity of the Anthropocene has informed a less bipolar and more sober (if no less ethically and politically committed) ethos of description in recent ecocriticism: of designing words, stories, images, concepts, and affects through which literature and its multiple environments can be captured. In this research context, an emphasis on *form* can help us make a case for the enduring relevance of literature in debates over the environment. Literary *form* can enrich interdisciplinary discussions by providing patterns, connections, structures, and descriptions that other kinds of knowledge production are less free to generate, if only because their protocols don't allow the blend of imaginative, speculative, and descriptive elements that makes up literary form. In first-wave ecocriticism, form *expresses* nature; in second-wave ecocriticism, form is something that also *shapes* and *deforms* nature; today, form can be a heuristic device that helps us come to terms with a planetary derangement we have only begun to understand and respond to.

The literary critical landscape today looks very different from the one in which ecocriticism initially emerged in the early 1990s. At that time, ecocriticism understandably set itself apart from approaches that were excessively invested in theoretical reflection, the critique of power, and the recognition of sexual, gender, or ethnic difference. It only later realized that all these foci could enrich rather than obscure our understanding of the relation between literature and the environment. Such an enriched and pluralized ecocriticism participates in two recent developments in literary criticism that have shifted it toward a more *descriptive* and a more *interdisciplinary* approach. Criticism has in recent years seen a notable shift from demystifying practices of critique to less suspicious critical postures that engage with texts in more nurturing and textured ways. For a critic like Rita Felski (2015), the "mood and method" of critique, often inspired by psychoanalysis, Marxism, and poststructuralism, is one of suspicion (1), which leaves unaddressed the myriad other tonalities in which literary works address us. Notions such as "postcritique," "surface reading," or "just reading" (Best and Marcus 2009) point to less heady and more descriptive reading practices that work hard to figure out what happens on (rather than below) the surface of texts. Description is not as boring as it sounds: when dealing with demanding and unruly objects such as literary texts and Anthropocene worlds, description involves an exploratory, imaginative, and creative dimension. Description is hard to get right, and the process of description is often an occasion to register the limits of existing formats and to reach beyond them. As such, description is "well suited to emergent evidence that exceeds but might ultimately be essential to reformulating the frame of analysis" (Marcus et al. 2016, 4). This is precisely what, in the previous section, we saw *Styrofoam*'s formal experiments accomplish in their attempt to give shape to the as-yet-undescribed Anthropocene life of thermoplastics. An intervention like Reilly's makes knowledge about this sprawling reality available for interdisciplinary dialogue; it is marked by an "essential generosity" pertaining to "the collective, uncertain, and ongoing activity of trying to get a handle on the world" (Marcus et al. 2016, 4). As an attempt to *describe* a complex world, literary form can make a crucial contribution to our understanding of the environment.

The second recent development that makes form a crucial resource is the rise of interdisciplinary research domains, which sees the study of literature liaise with other fields in new constellations such as the medical humanities, the digital humanities, or, in the case of ecocriticism, the environmental humanities. Environmental criticism's enlarged understanding of the digital, technological, urban, toxic, microbial, and transcorporeal environments we inhabit makes interdisciplinary dialogue necessary, as a measure of scientific literacy and a knowledge of other humanist disciplines (environmental history, political ecology, cultural geography) is required to get a grip on the multiple objects that environmental literature engages. The need for a broad conceptual repertoire for describing a weirded world has even informed a return of high theory. The deconstruction of Jacques Derrida and Paul de Man, which had its heyday in the 1970s and 1980s, is traditionally taken to exemplify everything that ecocriticism opposed: a wall-to-wall textualism, a cynical denial of literature's capacity to reflect its real-world referents, and a nihilist disbelief in social change and political action. I already pointed to the surprising renewed relevance of Derrida's "Il n'y a pas de hors-texte" ("there is no outside of the text") in my introduction, and in the work of critics like Timothy Clark, Tom Cohen, Claire Colebrook, J. Hillis Miller, and Timothy Morton, deconstruction's emphasis on incalculable complexity, uncertain authorship, the instability of humanist meaning, and the threat of unreadability has made it a rich resource for apprehending a climate-changed world (Clark 2010). Paul de Man tended to present deconstruction as a form of surface reading *avant la lettre*, or what he called "mere reading": a sustained focus on textual surfaces that aims to register textual operations without imposing aesthetic, ethical, or theological preconceptions (de Man 1986, 24). This explains how theory, like science, like many humanities disciplines, and like literature, can encounter other forms of knowledge in a shared project of *interdisciplinary description*. In the next section, I show how this ambition provides an accurate characterization of the environmental humanities, the field that has increasingly come to enfold ecocriticism.

Differentiating Difference: Form and Environmental Humanities

One remarkable feature of Reilly's *Styrofoam* is that it relentlessly focuses on an object—or, to use Timothy Morton's term, a sprawling "hyperobject"—that falls outside the customary purview of literary studies; Styrofoam is *different*, but not in the way differences that academic discussions typically have come to valorize are different: difference in gender, sexuality, ethnicity, and also (if often somewhat more reluctantly) in socioeconomic status. As Ursula Heise (2016) has noted, there is an irreducible tension between science's aspirations toward universal validity and "disciplines such as anthropology, history, or comparative literature, which have traditionally specialized in tracing differences between moments in time, communities, cultures, and aesthetic forms" (220). Even if the challenges of the Anthropocene require a concerted global approach, and even if they make it necessary to think planetary connectedness, Heise (2014) underlines that "[t]here is no freeway from ecological crisis to human universalism that does not have to retrace the byways and detours of difference."

Think, for instance, of the hype surrounding geoengineering—deliberate large-scale modifications to the earth system to address the unwelcome side-effects of human action. Plans to, for instance, add sunlight reflecting aerosol to the stratosphere or seed the oceans with iron fillings to increase their capacity to absorb carbon dioxide are marked by "a heavy-handed interventionist bias" that accustoms audiences to the desirability of techno-fixes (Emmett and Nye 2017, 86). In the face of such ambitions, an environmental humanities perspective can insist on, for instance, the challenges for global governance when one country's voluntarism clashes with the agenda of others; the blindness of such designs to the increased likelihood of famines and droughts in the Global South (Grove 2019, 38); or the insight that Maori, Samoans, Laplanders, and other ethnic groups have no desire to live in a deliberately re-engineered climate (Emmett and Nye 2017, 89). Environmental humanities approaches here join hands with the arts: there are clear echoes between a blockbuster film like *Snowpiercer* and a fictional work of popular science like *The Collapse of Western Civilization*, which I already mentioned in the introductory chapter. Bong Joon-Ho's 2014 film sees the last remaining humans survive in a high-speed train circling an earth that froze after a disastrous attempt to insert an artificial cooling substance into the atmosphere (Streeby 2018, 1–4). Oreskes and Conway's (2014) book imagines a failed "International Aerosol Injection Climate Engineering Project" that initially seems to succeed in cooling the planet, until it sets off the shutdown of the Indian monsoon, which then generates crop failures and famines that compromise political support for the project. This, in its turn, leads to an inadvertent cessation of the project, which informs a "termination shock" through which the planetary situation spins totally out of control (27–28).

Notice what Oreskes and Conway are *not* doing: they are not simply opposing all scientific and technological ambition, nor are they setting themselves up as the ethically attuned critical consciousness that tech bros are supposedly missing. Only a few pages later, they imagine an equally hubristic geoengineering project involving a pitch-black lichenized fungus that, later in the century, does manage to turn global warming around (32–33). What they do is emphasize that one crucial step in the deterioration of the planet is the withdrawal of Indian political support: the process involves, in other words, an interplay of different differences—technological, atmospheric, as well as cultural and political. Such a constructive, co-creative, and properly critical approach, and not a merely anti-scientific and anti-technological attitude, characterizes the environmental humanities at its best. A recent introduction to the field make this emphasis abundantly clear, as it insists that "the humanities must offer constructive knowledge as well as criticism" informed by an "understanding of the urgent need to take knowledge-driven action" emerging through "creative cooperation between the humanities and the sciences" (Emmett and Nye 2017, 2, 7, 9). Oreskes and Conway are not boasting that humanities perspectives are less deluded than scientific ones; they are adding humanities insights to provide a more salient, multifaceted, and ultimately more accurate description of environmental crisis.

The label "environmental humanities" is a loose one, even if it has become increasingly institutionalized since the beginning of the century and can be expected to become more firmly entrenched in the next decade (Emmett and Nye 2017, 6). For Ursula Heise (2014), it refers to "an intellectual framework that prioritizes connections between the various humanities disciplines that have pursued environmentally oriented research over the last few decades." For such research, collaborations with the sciences are indispensable, but attention to divergent cultures, histories, and values is equally essential for grasping the very nature of ecological crises. The contexts in which crises take shape are inevitably as much historical and sociocultural as techno-scientific. This constructive engagement with scientific knowledge echoes Bruno Latour's much-cited 2004 essay "Why Has Critique Run out of Steam?" In this essay, Latour worries that his earlier social critique of scientific practice is uncomfortably close to the arguments of climate change deniers and conspiracy theorists. Instead of his earlier uncompromising critical stance, Latour (2004) calls for "a *stubbornly realist attitude*," "another powerful descriptive tool" that makes environmental threat emerge as a "matter of concern" rather than a mere "matter of fact" (213–32). In line with this intervention, the environmental humanities can be considered as a cluster of perspectives that liaise with science to articulate the challenges of the Anthropocene as matters of concern.

Literary form—that is, the way literature shapes and organizes reality in ways that go beyond the disciplinary constraints of academic enquiry—is an important resource for such a project of interdisciplinary description. Attending to form shows how literature can foster knowledge that articulates cultural and historical difference with even more radical differences—between the human and the nonhuman, or between life and nonlife, for instance—*without* collapsing the difference between these differences. Literature's formal affordances offer ways to engage with the ongoing reorganization of life and the relation between life and nonlife. If the notion of life has always been a complex one—philosopher Eugene Thacker (2010) has called it awkwardly "human-centered and yet unhuman-oriented" (ix)—recent developments have further destabilized it: think of the upscaling of human life to the status of a geological force; projects in species revivalism that promise to extend particular life forms beyond their expiry date; the intrusive re-engineering of biotic material, as when deep-sea microbial genes are turned into corn to make ethanol; forms of artificial life that decouple the logic of life from its organic basis and simulate life as pure digital information; or the discovery of lateral gene transfers—genes leaping among contemporaries rather than being transmitted to the next generation—in marine environments that uproot the Darwinian tree of life by exploding the discreteness of species (Helmreich 2011). All these developments require new imaginaries, concepts, and narratives, and literature is one place where these can emerge. They also provide an impetus for literary innovation. As Stephanie LeMenager (2017b) remarks, artistic formats are "fraying, recombining, or otherwise moving outside of our expectations of what they ought to be because life itself is moving outside of our expectations of what it ought to be. It is worth considering how life itself begins to encourage new representational regimes" (477). Tracing such regime change, then, requires a commitment to form.

When nature is no longer a stable background for human culture, and when life unpredictably shuttles between natural and cultural poles, life and form enter into novel constellations. In order to capture the interactions and overlaps between biological and cultural dimensions, anthropologist Stefan Helmreich introduces a helpful distinction between *life forms* and *forms of life*. Life forms, for Helmreich (2009), are biological entities; they are "embodied bits of vitality called *organisms*," and the ways in which organisms interact with their surrounding ecologies and with one another (6). Forms of life, in contrast, refer to "those cultural, social, symbolic, and pragmatic ways of thinking and acting that organize human communities" (6). Forms of life always form a plurality of forms of an "uneven, contested, and overlapping character" (6). In a time of biotechnology, genomics, species revivalism, and geological agency, the relation between life forms and forms of life, Helmreich writes, has become "liquid, turbulent" (8). A phenomenon like species revivalism, for instance, is as much a question of the genetic engineering of a particular life form as of the meanings that human forms of life ascribe to biodiversity, historical responsibility, and extinction—perhaps even the fear of the human's own extinction. Species revivalism also impacts the fate of different life forms: if, for instance, de-extinction biologists were to reanimate the passenger pigeon, a North American species that went extinct when Martha, the last known living specimen, died in 1914, these revived birds would encounter forests dominated by red oaks, which over the last century grew at the expense of the white oaks in which the passenger pigeon found abundant tree fruits (Emmett and Nye 2017, 73–74). To compound the complexity of the entanglements between life forms and forms of life, Thom van Dooren (2014) has remarked that the notion of a form of life not only applies to humans: there are also other-than-human forms of life, as he considers "birds (and other organisms) as life forms with a form or way of life" (8–9). In an age when life forms are loosened from biological substance and forms of life are extended beyond the human, a better understanding of the relations between life and form is an urgent matter of concern. It is at this juncture that literature has its job cut out for it.

Environmental Humanities at Sea: The Ocean and the Life of Form

Reilly's *Styrofoam* renders the ocean as a nonhuman place, taken over by what poet John Ashbery, in his blurb for the collection, calls "[a] vast Sargasso sea of plastic fragments the size of a continent." Traditionally, the ocean has served as the backdrop for imperial conquest and colonial adventure. The clearest emblem of that worldview is the frontispiece to Francis Bacon's 1620 *Instauratio Magna*, which contains the tract *Novum Organon*, a text that theorizes the role of science in the Western conquest of the globe. The frontispiece depicts a ship sailing between the Pillars of Hercules (the old name of the promontories flanking the Strait of Gibraltar), then considered the limit of the known world. In the distance, another ship has set sail for the undiscovered New World. In the modern imagination, the ocean figures as a flat surface to be crossed, as an "inert backdrop" rather than as "wet matter" (DeLoughrey 2019, 22). This erasure of the material reality of the ocean continued when the humanities began to pay attention to the ocean near the end of the twentieth century. Paul Gilroy's influential 1993 book *The Black Atlantic* showed how attention to the ocean could help the humanities destabilize national or continent-bound imaginaries. Gilroy (1995) refashions the Atlantic as a zone of hybridization and intercultural affiliation in which the afterlives of the Atlantic slave trade have generated vibrant and convivial cultural practices and forms. Since the turn of the century, studies of the cultural and social histories of the ocean have gone under the name of oceanic studies, blue cultural studies, the new thalassology, or the maritime humanities. The ocean emerges in these studies as a site for novel forms of relationality—of what Hester Blum (2014), a leading critic in the field, has called "erosion, drift, dispersion, confluence, solvency" (35). Elizabeth DeLoughrey, another important voice, notes that the emphasis has been on "concepts of fluidity, flow, routes, and mobility" in what is essentially a "*transoceanic* ... narrative of flat surfaces rather than immersions" (2019, 22; 2017, 33).

Only more recently, forms of "critical ocean studies" and "hydro-criticism" (DeLoughrey 2017; Winkiel 2019) have taken on board the realization that the ocean is not just a surface that can be traveled, mapped, and navigated, but also a vast volume that resists human control and alters the parameters of planetary life. The differences the oceans deliver are not only cultural but also *ontological*, as the deep sea houses life forms that upset biological taxonomies or invites us to think of space as a dynamic force rather than a place (Steinberg 2013). And even if the sea is the main infrastructure of capitalist globalization (over 60 percent of the global oil supply is transported over sea), the depths of the ocean remind us that what we inhabit is less a globe we can map and control than a *planet* that resists human designs (Wenzel 2014, 21; more on the distinction between planet and globe in the next chapter). Stacy Alaimo (2016) thinks of the ocean as such a planetary space: "The depths of the ocean," she writes, "resist flat terrestrial maps that position humans as disengaged spectators" (161). Skimming the surface will no longer do, as "[t]o begin to glimpse the seas, one must descend, rather than transcend": marine scientists, in other words, "must ... become submerged" (161).

Submergence is the title of a 2011 novel by the Scottish writer J.M. Ledgard. Deeply informed by scientific, technological, and humanistic knowledge, the novel invites dialogue with the environmental humanities and critical ocean studies. Importantly, it does so on unabashedly *literary* terms: its intervention in current thinking about the ocean relies on literary form to articulate unexpected alliances of forms of life and life forms. In an interview, Ledgard qualifies his novel, which sounds human and nonhuman dimensions of the ocean, as a form of "planetary writing" rather than "nature writing." What makes the ocean a promising site for moving beyond human constraints is, Ledgard notes, that it is "confounding":

> you have no breath in it, no light, and consequently no imaginable human life, yet it is immutable, and when you stack it up you find it is nearly all of the living space on the planet. What I wanted to do was to alter the reader's perspective of earth, to show that … to step out on a field is rare while to float and scintillate with bioluminescence is common.
>
> (Gourevitch 2013)

Ledgard's novel draws on the ocean as a site of *defamiliarization*: it allows his novel to confound readers' expectations and provide new apprehensions of reality. Interestingly, Viktor Shklovsky's 1917 essay "Art as Technique," the very text that codified the idea that literature is a technology of defamiliarization, links the operation of literature directly to form and to the confrontation between life forms and forms of life. Shklovsky (1965) defines defamiliarization as the work of form: to make things unfamiliar is "to make forms difficult," to the extent that defamiliarization "is found almost everywhere form is found" (12, 18). One of Shklovsky's main examples is Leo Tolstoy's story "Kholmstomer," which is narrated from the perspective of a horse. Ledgard's debut novel, *Giraffe* (2006), is partly told from the perspective of a giraffe and describes a freak historical incident in which the Czechoslovakian secret police in 1975 massacred a group of 49 giraffes for no discernible reason. This defamiliarizing perspective destabilizes the relation between different life forms (human and giraffe) as well between different forms of life, such as the animals' captivity and the mental confinements of the communist mindset.

Submergence stages the estrangement of customary forms of life by different life forms in more daring ways. The novel presents a fairly traditional (and very human) love story that brings together Danny, a colored female biomathematician working on deep ocean life, and James, a British spy working in Africa and posing as a water engineer who, at the outset of the novel, is captured by a group of Somalian Jihadis. The templates of the romance and the spy thriller are familiar enough, but these human stories constitute only two of the novel's strands: these story lines are juxtaposed with a collage of materials pertaining to the science and cultural history of deep sea life. Danny devotes her life to the study of microbial deep ocean life: life that persists in the darkness, without photosynthesis, and that in the history of terrestrial life precedes the animal and plant life that later "ascended from the eternal night … heading towards the light" (Ledgard 2012, 95). The novel underlines that bacterial life continues to exceed other forms of life both in terms of biomass and genetic diversity. Knowledge of this life form, Danny notes, is necessary "to comprehend the scale of life on earth … The fact that life can exist in the darkness, on chemicals, changes our understanding about life everywhere else in the universe" (137).

The novel's recurring evocations of the facts and fantasies of deep ocean life submerge human life in "the pullulating life in the dark parts of the planet" (9), and this means that it radically estranges customary ways of looking at human forms of life—as a spy plot, as a love affair: "[t]he microbioal life of the deep," the novel notes, contains "fleeces of microbes … that are even more extraordinary than those that live on the timbers of our eyelashes" (156). The maritime world, in *Submergence*, is many things simultaneously: a site of human endeavor and strife (think of the Somalian pirates); an archive of human violence (Danny is haunted by the image of a sinking slave ship carrying one of her ancestors); a site of encounter (the love affair is situated at a seaside hotel); but also a radically "alien ocean" (to appropriate Stefan Helmreich's phrase) that the most advanced life sciences have only begun to sound and whose impact on human life Ledgard's novel is one of the first literary works to explore. *Submergence* acknowledges the modern novel's insistent fascination with the sea as an area of adventure, travel, and contemplation, only to note that the novel's concern has mainly been with beaches and tides, while "the connection with the ocean has been lost" (95); the history of the novel has tended to forget that "there is so much darkness in our world" (95). Jules Verne's *Twenty Thousand Leagues under the Sea*, for instance, refers in its title "to the journey taken across, not down" (96). The downward journey to a planetary darkness is then *Submergence*'s contribution to this literary tradition.

Ledgard's novel is attentive to the cultural implications of scientific insights as well as to the ways scientific enquiry is framed by geopolitical and aesthetic concerns. But its contribution to the environmental humanities is not only thematic. Crucially, Danny's knowledge of the deep disconnects her from James: "She could not simplify the maths for him" (112). By juxtaposing, rather than seamlessly integrating, two all too human plotlines with the myths, legends, anecdotes, factoids, and fantasies pertaining to deep sea life, the novel presents the relation between human forms of life and different life forms as a kind of *entanglement without intimacy*, as a type of relatedness that remains marked by unfamiliarity, alienation, power differentials, and missed connections. *Submergence* uses literary form to describe the ocean in a way that complements and corrects oceanic studies' emphasis on drift and dispersion, on flows and fluidity; its focus on the dark side of the ocean not only resonates with Reilly's *Styrofoam*, but also with more recent interdisciplinary debates that have begun to factor in the irreducible materiality of the ocean.

The novel's collage aesthetic shows that different forms of life and life forms do not necessarily become hybrids or get attached to one another or the places they inhabit; instead of such an emphasis on local attachment and on what Ursula Heise (2008) has called an "ethic of proximity" (33), certain life forms remain withdrawn from cultural forms of life: there is still a "division between life on the surface of the world and the life … in the Hadal deep" (Ledgard 2012, 139), just as there is no understanding between the Somalian Jihadis and the Westerners they meet. The life below is presented as a limit to literature's form-giving capacities; it is referred to as the life of "teeming hordes of nameless micro-organisms that mimic no forms, because they are the foundation of all forms" (180). The novel recounts the story of the late-nineteenth-century HMS Challenger, whose job it was "to plumb the remote seas and trawl them for new life" (97). The crew managed to bring up tens of thousands of new species, as well as seemingly insignificant heaps of slime. What they did not know was that this slime was also a remainder of life: "the slime which covered the inside of the dredge each time it was brought up was not the unexceptional ooze the ship's scientists believed it to be. Not whale snot, either. It was all that remained of the most exquisite forms of millions of sea squirts, salp, and jellies, whose diaphanous musculature—more remarkable than any alien species yet conceived—had lost its form in air" (97). This episode shows that the relations between life, nonlife, and form are intricate, mutable, and unpredictable—and therefore in need of the complex acts of description that literature can provide.

Submergence's interest in geopolitical and biological themes resonates with the increasing concern for cosmopolitan connectedness and material agency in what I earlier identified as the third and fourth waves in ecocriticism. Like Reilly's *Styrofoam*, it participates in a project of interdisciplinary description that is increasingly institutionalized under the rubric of the environmental humanities. The last point I want to make in this chapter is that the environmental humanities are not only a project of knowledge-oriented description; they are, in fact, often closely allied to environmental activism and foster collaborations *beyond* the university and the literary field. Ledgard's career since the publication of *Submergence* illustrates these multiple alliances very well, as it takes the lessons learned from the ocean back to earth and ultimately to the sky. Between 2012 and 2016, Ledgard, who earlier worked as a frontline foreign correspondent for the *Economist*, led Afrotech, a Lausanne-based consortium of leading roboticists, architects, engineers, designers, and logisticians that seeks to build the first droneport in the world in Africa. The idea is to develop a network that allows cargo drones to deliver healthcare and supplies (but also commodities) to off-grid communities that, because of failing transportation infrastructure and environmental challenges, are currently locked out of trade circuits. If the spread of cell phones has made it crystal clear to (overwhelmingly young) African populations that they are missing out on massive opportunities, drone delivery systems can help meet their aspirations and contain global unrest by transporting the commodities these populations crave.

However cynical this may sound, Ledgard's (2015) latest publication, entitled *Terra Firma Triptych*, illustrates the value of humanities perspectives for such efforts to manage a crowded and unequal Anthropocene world. The triptych contains three texts: two idiosyncratic travel stories and a kind of manifesto for the drone initiative. Ledgard's embrace of drone technology is informed by the realization that "the extractive nature of colonial and postcolonial rule," which only needed roads to transport raw materials from mountain to coast, has made Africa's road system useless for medium-distance trade (loc 272–74). The manifesto compares the "Red Liners" to an albatross (loc 282), and as in Reilly's *Styrofoam*, this reference serves as a recognition of the dangers of human overreach. The drone initiative, Ledgard underlines, is not a form of "techno-utopianism" (loc 351), and he notes that human apprehensions about unmanned flight and robotification need to be acknowledged: "The biggest hurdle to the mass adoption of cargo drones and droneports is human emotion," so "it makes sense to think carefully about how they should look and feel" (loc 358, loc 364)—imitating birds, for instance, or mimicking the drift of an octopus underwater. Again, cultural forms of life and biological life forms enter a complicated choreography that enriches a technocentric approach.

Developing drone networks is a way of *managing* capitalism and populations; it enhances the world that exists and does not imagine an alternative to it. The manifesto is preceded by two strange short travel narratives, set in Sudan and Rwanda respectively, that anticipate the critique that Ledgard's intervention is insufficiently radical. Both stories ponder the desire for a part of nature that is not spoiled by human interference, for "the absence of human settlement of any kind" (loc 39), only to conclude that such repose is no longer available. What, in the Sudan story, seems to be a perfectly flat *terra firma* ("solid land") is actually shot through by burrows, aardvark holes, and tunnels "perforating the earth" (loc 128). And what, in Rwanda, initially looks like a valley that "had become unaccustomed to humans" (loc 196) after the genocide in the 1990s, turns out to be a site vacated for a planned international airport. What appears to be a solid alternative to the liquid depths Ledgard explored in *Submergence* turns out to be no such thing, and what looks like a site of splendid isolation is so only in the quiet before a manmade storm. In such an unstable and thoroughly compromised world, Ledgard seems to suggest, alleviating frustration and unhappiness in disadvantaged populations is a worthy goal.

The traveler in the Rwanda story imagines the future of the vacated site as a connectome, a neuroscientific map "of all possible neural connections in the brain and nervous system— pulsing, roaring, getting denser" (loc 226). On the face of it, this reads like a clichéd metaphor for a networked, hyperconnected world. But in the endnotes, Ledgard underlines that a connectome is not a lazy commonplace, but a phenomenon that the life sciences are only now beginning to get a grasp on. The connectome, then, does not offer a glib vision of frictionless interconnectedness (something *Submergence* also resisted); instead, the image foregrounds the hard work of articulation and collaboration, of imagination and description. As the story notes: "It was not good, neither was it awful" (loc 235). It is, quite simply, the Anthropocene present, and Ledgard's attempt to bring his interdisciplinary intervention to bear on the present aligns it with the larger project of the environmental humanities. If this chapter has foregrounded literature's ability to make us describe and understand the present, the next chapter is interested in how it makes us feel and inhabit it.

2 Genres, Media, Worlds

Deranged Realism: Genre and Affect

One reason for environmental thought to turn to literature is literature's capacity to capture the reality of the natural world. When, in the Anthropocene, that reality turns out to be weirder and unrulier than we had assumed, literature's vaunted realism comes to seem strangely *un*realistic. Ecocriticism traditionally valorized literature's *mimetic* powers: its ability to represent the natural world in reliable and transparent ways. In the early stages of the field, this emphasis on the referential and experiential dimensions of literature informed a preference for poetry and nonfiction at the expense of, for instance, the genre of the novel. And when second-wave ecocriticism later discovered the novel, it tended to prefer traditional realist modes at the expense of more difficult, experimental, modernist styles.

Yet realism is not just a neutral mode of notation that opens an undistorted window onto the world out there: like all modes of representation and storytelling, it inevitably encodes particular values and preconceptions. Nineteenth-century realism's pedantic attention to the material minutiae of everyday life, for instance, is often taken to reflect as well as reinforce a bourgeois mentality (Moretti 2007), while the realist novel's traditional focus on the ways individuals negotiate a tenuous balance with the social world cements a particular idea of what it means to be a well-adjusted subject (Armstrong 2005). Another way of putting this is to say that realism is a *genre*: a set of historically specific formal and thematic conventions that imply values and meanings. By keeping these commitments implicit and claiming to present a faithful image of the world, realism signals to reader that these values are self-evident, not even in need of being made explicit; in this way, realism is a powerful tool for naturalizing a particular worldview.

The Anthropocene unsettles long-held convictions about the world and disables realism's capacity to pass off those convictions as common sense. One of the most frequently voiced ideas about literature and the Anthropocene is that the ebbing rhythms and moderate scales of realistic fiction are fatally out of sync with the new realities of the Anthropocene. The most famous instance of this critique is novelist Amitav Ghosh's 2016 book *The Great Derangement: Climate Change and the Unthinkable*. For Ghosh, the modern novel is unfit to render our deranged world, for at least three reasons. First, novelistic realism constrains the imagination to the realm of *the probable* and the credible; the mundane lives of novel characters cannot be disturbed by freak weather events, alien invasions, or sudden extinction events without works losing their standing as serious literature. Realism remains blind to the frankly improbable ways in which the Anthropocene derangement of Holocene conditions proliferates uncertainty and risk, as these distortions are "not easily accommodated in the deliberately prosaic world of serious prose fiction" (26). Besides, realism limits itself to the domain of *the human*, and fails to factor in the agency of nonhuman forces (and this is a second constraint): "it was in exactly the period in which human activity was changing the earth's atmosphere that the literary imagination became radically centered on the human" and banished the nonhuman to the domains of science fiction and fantasy (66). With the focus on probability and the human comes a third limitation—one of *scale*: the novel, Ghosh notes, invests its realist energy in localized settings, in the evocation of particular and necessarily limited places and periods, and is thus incapable of tracking the larger continuities that define the Anthropocene present:

> the earth of the Anthropocene is precisely a world of insistent, inescapable continuities, animated by forces that are nothing if not inconceivably vast. The waters that are invading the Sundarbans are also swamping Miami Beach; deserts are advancing in China as well as Peru; wildfires are intensifying in Australia as well as Texas and Canada.
>
> (62)

Ghosh's diagnosis of the limitations of realism is less controversial than the note of defeat on which he ends his analysis. His confident prediction that serious fiction "will double down on its current sense of itself" (71) is strangely uninformed by recent developments in the literary field. At the same time, he remains blind to the capacity of forms of genre fiction—

fiction that, unlike serious literary realism, wears its reliance on generic patterns on its sleeve—to accommodate the vaster and more dispersed agencies that Ghosh finds banished from the realist novel. Mark McGurl (2012) has shown that "the spatiotemporal vastness and numerousness of the nonhuman world" (537) has found expression in the genres of science fiction and horror. Unlike the literary novel, which must uphold its seriousness, these genres are "willing to risk artistic ludicrousness in their representation of the inhumanly large and long" (539). For Eugene Thacker (2011), similarly, the genres of science fiction and supernatural horror can unsettle the comforts of the world we think we know, and evoke a sensation of horror that, he argues, counts as "a way of thinking the world as unthinkable, and the limits of our place within that world" (98–99). In a less mystical register (and Thacker explicitly links horror to mystical traditions!), Ursula Heise (2018) has noted that "science fiction is becoming the default genre for the narrative engagement with climate change." Because of the genre's ability to reimagine the present as the past of a future yet to come, science fiction is less constrained by the demands of probability and the human scale, and can address urgent present-day concerns such as the dangers of biotech and energy depletion through the detour of, for instance, an imagined future Thailand (this is the set-up of Paolo Bacigalupi's 2009 novel *The Windup Girl*).

Rather than bickering over the alleged superiority of realist or genre fiction, or of serious or (in McGurl's term) "ludicrous" works, I want to pick up two more productive points from this discussion. The first is that genres are never stable and rigid sets of prescriptions (a sort of recipe for baking literature), but rather function as flexible templates that, moreover, intermix with one another within particular works (they are also the *ingredients* of literary works). *The Windup Girl*, for instance, combines science fiction world-building with a fast-paced thriller plot; we already saw in the introduction that a novel like *Annihilation* "weirds" genres such as gothic and horror (two genres that themselves have overlapping features). Genres, that is, are flexible means to a literary end, whether that end is a cognitive mapping of the present (something to which science fiction and realism are particularly conducive), the production of a certain sensation in the reader (for which horror and the weird are particularly apt), or the expression of a particular world view. The imaginative challenge of the Anthropocene, then, does not have to result in a doubling down or breakdown of generic conventions, as Ghosh would have it; it has often resulted in a recombination of existing genres and a transformation of existing templates. In the first book-length study of Anthropocene literature, Adam Trexler (2015) observes that "climate change necessarily transforms generic conventions" (14), as it makes science fiction's customary techno-optimism implausible, or makes well-intentioned realist depictions of everyday life read like satires of consumerism. Already in 2005, Ursula Heise noted that "[e]cological storytelling frequently relies on and transforms traditional literary genres" (Heise 2005, 129), especially as it sees itself confronted with planetary connections and time-scales that exceed those of individual lives. The phenomenon of "cli-fi," a term that refers to literary works that explicitly deal with climate change, can be understood as such an assemblage of different interlocking templates, rather than as a new discrete genre in its own right. Indeed, a neat separation between different genres is an increasingly marginal feature of the contemporary literary field, as the migration of genre elements into the literary mainstream (in the work of prominent novelists like Margaret Atwood, Junot Díaz, Jennifer Egan, and Colson Whitehead, for instance) has by now become entirely routine.

A second point concerns the central importance of affect in discussions over Anthropocene genres. We encounter affect in notions such as "ludicrousness" (McGurl), horror (Thacker), and the weird (VanderMeer), but they also surface in Ghosh's (2016) assessment on the realist novel's redundancy, when he notes that "climate change events" are "too grotesque, too dangerous, and too accusatory to be written about in a lyrical, or elegiac, or romantic vein" (32). Anthropocene affects, it appears, are too excessive and too disorienting to be captured by the emotive protocols of traditional literature. This explains why Anthropocene literature warps, bends, and recombines different generic templates. Genre and affect are intimately connected, as recognizing generic templates cues readers into expecting particular feelings. For Lauren Berlant (2008a), genre is "an aesthetic structure of affective expectation" (4), it is "a loose affectual contract that predicts the form that an aesthetic transaction will take" (Berlant 2008b, 847). Anthropocene literature's experiments with genre are then efforts to rewrite the terms of that contract, or to deliberately frustrate the expectations that stale generic conventions generate. Think of *Annihilation* again: VanderMeer's novel sets up the expectation that it will deliver science fiction (it is set in an undefined near future, populated by scientists) and horror (we soon learn that many people have inexplicably disappeared), yet it ends up warping these expectations and delivering something altogether more … weird. For Roger Luckhurst (2017), the weird "inheres in perversity and transgression. It twists or veers away from familiar frames and binary distributions" (1052), and therefore also from predictable, pre-programmed emotions. Anthropocene fiction's genre-bending, then, often produces inscrutable, quaint, and awkward affects, rather than the more grandiose emotions of traditional literature (think of sympathy, exaltation, or catharsis).

Environmental literary studies has not unequivocally embraced such hesitation and unpredictability. After all, such uncertainty about the impact of literary form on the reader and about the translation of that psychological impact into real-world action conflicts with what Alexa Weik von Mossner (2016) has referred to as ecocriticism's fundamental trust in "the ability of environmental narratives to have lasting effects on the attitudes and behavior of readers" (534). According to Weik von Mossner, "there seems to be a certain consensus that affect and emotion play an important role in such processes of change" (535). Weik van Mossner herself is a leading voice in the interdisciplinary campaign to provide the study of environmental narratives and their influence on their audiences with some solid empirical grounding (a project she calls "empirical ecocriticism"), and to investigate "how storytelling practices themselves find their place in broader ecologies of action and interaction" (Lehtimäki 2013, 121). Weik von Mossner draws on insights from cognitive science, affective narrative theory, and the psychology of fiction to explore the psychological impact of reading fictional actions and nature descriptions. Weik von Mossner is a firm believer in the workings of empathy: when we read about a character losing her footing, this leads us to "a *personal* version of that experience" (539). The sensation that fiction represents is affectively re-experienced by the reader. Weik von Mossner calls this procedure "the mechanism of strategic authorial empathizing" (2016, 546), a notion that betrays a strong commitment to the agency of the author, arguably at the expense of other forces involved in the production of textual effects. One such complicating influence she underestimates is that of different generic templates. For affect to travel reliably from author to reader, narratives need to be immersive, and, Weik von Mossner writes, "some narratives are … more immersive than others, and the genre conventions of the realist novel are the ones that most deliberately and most consistently seek to ensure an immersive reading experience for the reader" (543). It is not clear what Weik von Mossner would make of readers who are horrified, overwhelmed, alienated, or baffled by texts they fail to immerse themselves into.

The econarratological work of Erin James (2015) might seem to leave more room for divergent experiences. As narrative comprehension requires "mental modeling" as well as "emotional inhabitation" (33), James believes that readerly immersion opens the gate to a better understanding of different perspectives on environmental issues. Environmental storytelling, James writes, "help[s] us understand the environment from the perspective of others, and thus experience the world according to alternative environmental imaginations," which "can help bridge imaginative gaps" and thus "can have important real-world consequences" (2–3). While James invokes insights from cognitive narratology to investigate how certain narrative affordances affect their readers, the wishful repetition of the word "can" in these quotations (and of repeated and unhelpfully vague formulations noting that mental simulation and genuine experience are "linked intimately" [19]) underlines that such scientifically grounded perspectives only *minimize* but don't *close* the gap between the literary text and its audience. When trying to describe genre's affective operation, a measure of indeterminacy inevitably remains.

This inexorable moment of tension between generic expectation and affective effect is strangely appropriate for the deranged world of the Anthropocene. As I explain in the third section of this chapter, the Anthropocene alters the relations between earth and world, between globe and planet. These relations are beset by frictions, gaps, and fissures. A literature that wants to describe this new constellation will inevitably be marked by inconsistencies and tensions. As I showed in the previous chapter, the operations of literary form in the Anthropocene include an acknowledgement of forces that resist form, and it is precisely these resistant forces that emerge in and through the attempt to describe an unruly reality. An econarratological approach like James's glosses over such frictions as it promotes the worldly powers of fiction. Her case for fiction invokes the terms of a traditional liberal cosmopolitanism, as it sees environmental fiction "foster respect for comparison, difference, and subjectivity," which in its turn can "foster more sensitive and informed discussions" and "emotional connections" (208). Yet in the Anthropocene, the efficacy of such "ecoglobalist affects" (Buell 2007, 227) and such "eco-cosmopolitan" designs cannot be taken for granted, as notions such as *kosmos*, world, and globalization have begun to change their customary meanings.

In the next section, I illustrate the point that the Anthropocene destabilization of the earth disturbs literature's world-building capacities through a discussion of one cluster of texts that are intertextually related. I focus on Edgar Allan Poe's only novel, *The Narrative of Arthur Gordon Pym of Nantucket*, and its partial rewriting in two contemporary novels, Yann Martel's *Life of Pi* and Mat Johnson's *Pym*. These three works explore different forms of Anthropocene world-making: the island, the lifeboat, geoengineering, multispecies communities, and the biodome. All these forms aim to establish a self-contained microcosm, but these constructions all fail to exclude uncontainable planetary forces. In all three novels, the volatility of the world registers as generic instability. This shuttling between different generic template, in its turn, generates an uncertain affect. My point is not that we need to indulge in this uncertainty, but that it is an unavoidable dimension that emerges when we try to describe these texts' operations as accurately as possible. Patiently describing how genres, affects, life forms, and forms of life interlock in these narratives makes predicting reader responses and their translation into real-world action a very speculative affair. Still, it may ultimately tell us more about how literature records the Anthropocene world than empirical approaches, which downplay this speculative dimension, can. After my case study on Antarctic literature, the rest of this chapter teases out the tensions and overlaps between notions such as world, planet, earth, and globe (in the third section) and attends to the ways different media—literature, but also photography, cinema, and post-cinema—can access those tensions (in the fourth and final section). All of these media can intimate the strangeness of the Anthropocene, but they all do so in very specific ways.

Poe–Pi–Pym! Literature vs. Antarctica

Antarctica has always constituted a challenge to the literary imagination: it lacks an Indigenous population (and thus a homegrown literature), and the relentless blankness of the landscape's ice and snow is dispiriting at best. Antarctica's featureless wasteland has inspired "a history of negative discovery, a hermeneutics of despair" (Wilson 2004, 37). One imaginative strategy for coping with such despondency has been the projection of lurid images onto Antarctica's blank screen: the literature of Antarctica is marked by a proliferation of vortices and giant lodestars, of polar holes leading to an interior earth, sometimes all the way to the North Pole, and of "Lost race" fantasies (Leane 2016, 34, 46). Such sensational figures convert the Antarctic into an affective space, a space that conveys the darker aspects of the traditional sublime—obscurity, vastness, isolation—without quite allowing the human mind to recover from the traumatic impact of the encounter with the continent's nonhuman surfaces. Mariano Siskind has shown that the Antarctic continent has consistently rebuffed attempts at colonization and integration into a globalizing world (Siskind 2005); in a comparable way, literature about Antarctica has registered a radical limit to humans' (and their literatures') world-building capacities. A brief look at one particular intertextual archive points to the way affect, genre, life, and form have entertained this limit.

The Narrative of Arthur Gordon Pym of Nantucket (1838) is a novel of the early Anthropocene: its narrator sets out on a maritime adventure as a stowaway on a whaling ship, but after a mutiny and a shipwreck he enthusiastically joins the *Jane Guy*, the ship that rescues him, on an obsessive mission to map and cultivate the land and subdue the native population of Tsalal, an Antarctic isle. Unsurprisingly, the combination of scientific interest and colonial ambition, of exploration and exploitation, is reflected in the novel's obsession with race, and its almost hysterical insistence on clear contrasts between white and black. The Indigenous population's complexion is "jet black"; they are clothed "in skins of an unknown black animal" (Poe 1999, 163–64); in a startling detail, even their teeth turn out to be black (216). In one of the novel's many bewildering inconsistencies, the narrator's companion, Dirk Peters, is initially described as a monstrous creature "with an indentation on the crown (like that on the head of most negroes)" (49), before Poe decides to forget this portrayal and reinvent Peters as a white character. The novel ends with the narrator fleeing the insurgent natives—whose shrieks of "*Tekeli-li!*" fill the sky—on the "wide and desolate Antarctic Ocean" (211) and being miraculously saved by "a shrouded human figure" whose skin color, we read in the novel's very last words (before a long endnote that adds further confusion), "was of the perfect whiteness of the snow" (217).

Poe's baffling story activates several generic frames, none of them a traditionally realist one. The book announces itself as a sensational adventure story (the title page presents a breathless summary of the plot, culminating in "incredible adventures and discoveries STILL FARTHER SOUTH"); in the narrator's time as a stowaway hiding below deck, it exploits the claustrophobia of gothic fiction; after the shipwreck, it becomes a shipwreck narrative, including the horrors of cannibalism; setting sail farther south, it morphs into a diary-like combination of a scientific report (which Poe cobbled together from existing stories) and travelogue. In the end, none of these frames quite fit or last, and it is the mismatch between the different frames, together with the maddeningly unequal pacing of the narrative and the factual inaccuracies, that best conveys the sense of disorientation that disturbs the novel's worldbuilding. Neither the familiar emotive scenarios of trauma or the sublime quite come off: the perfectly white *deus ex machina* is surely sublime, but it is also quite ludicrous. Arthur Pym, for his part, is a remarkably untraumatized character (Wilson 2004, 40). He mentions his cannibalism matter-of-factly ("Let it suffice to say that … we devoured the rest of the body piecemeal" [117]) and insists that the horrors of shipwreck have not left him emotionally scarred: "The incidents are remembered, but not the feelings which the incidents elicited at the time of their occurrence" (136).

The different generic frames the novel tries on fail to program particular emotive scenarios, but this does not mean that the sensation of the mismatch between these templates conveys a robust sense of the geological realities of Antarctica. Indeed, the novel spectacularly misses the signs of geological agency it nevertheless intimates. Confronted with the irregular cavities and protuberances of a chasm on the Antarctic island, Arthur Pym makes a number of drawings of their structure (reproduced in the book) that resemble alphabetical characters, bringing the novel to speculate that "the hieroglyphical appearance was really the work of art," and to assign authorship to an ethnic group combining knowledge of Ethiopian, Egyptian, and Arabian characters (220). Far from a confrontation with nonhuman agency, then, this points to a failure to even imagine geological action.

This is even more apparent when geological events are captured in overtly racist terms. When the protagonist escapes from the isle in the midst of a volcanic eruption, the novel describes this geological phenomenon as a racial battle between black and white: the water is "of a milky consistency and hue" as "[t]he white ashy material fell now continually around [them], and in vast quantities" (215–16), until the mysterious white figures liberates them from "[a] sullen darkness" overtaking "the milky depths of the ocean" (217). Earlier on, the novel presents what may very well be a geological event as an unexpected attack by the natives. It notes that "the channel or bed of [the] gorge was entirely filled up with the chaotic ruins of more than a million tons of earth and stone," but rather than seeing this as a geological process, the novel aggressively overlays this interpretation with its racist imaginary as it concludes that the earth and stones "had been artificially tumbled within it" (187). "[A] partial rupture of the soil," the novel concludes, must be the work of the savages using cords to acquire "a vast leverage" (188). The certainty of black mendacity, for Poe, obscures the insight into geological agency, as such an insight would threaten the colonial posture of mastery Arthur Pym has come to adopt and for which the novel never quite finds the right genre.

Yann Martel's 2001 success novel *Life of Pi* remixes many of the elements of Poe's novel in its engagement with the Anthropocene, even if it abandons that novel's fascination with the South Pole. *Arthur Gordon Pym* prefigures the Anthropocene imagination by projecting an oblique and deeply colonial terraforming fantasy onto the natives; *Life of Pi* explores the possibilities and liabilities of a lifeboat scenario. The novel tells the story of Pi, an Indian boy who loses his parents in a shipwreck that leaves him stranded on a raft with the few remaining animals from his parents' zoo. After the death of a hyena, a zebra, and an orangutan, most of the novel is devoted to the multispecies cohabitation of Pi and the tiger Richard Parker. Richard Parker is the name of the cannibalism victim in Poe's novel (as well as of two [!] real-life cannibalism victims that postdate Poe's novel), while Tiger is the name of the dog accompanying Arthur Pym when hiding as a stowaway. Nor do the echoes of Poe's novel end there: apart from the two novels' elaborate metatextual frames, there is, for instance, the paradisiacal island on which Pi arrives and that turns out to be a massive carnivorous organism—"a free-floating organism, a ball of algae of leviathan proportions" (2001, 271–72)—which borrows the horror and treacherousness of Poe's Tsalal.

Life of Pi makes it very clear that the cohabitation of tiger and human being is not a starry-eyed return to nature, but a matter of careful life management. Indeed, Pi's painstaking chronicling of his interactions with the tiger recalls the novel's earlier celebration of the zoo as an institution that *liberates* animals from a life of "compulsion and necessity within an unforgiving social hierarchy" in the wild (16). Such careful management, the novel notes, is necessary to "*Keep Him* [the tiger] *Alive*" (166), and to keep the multispecies lifeboat—including microbes, bacteria, and "a multitude of sea life" (197)—afloat. As Eva Horn (2013) has shown, lifeboat imaginaries induce a situation of scarcity in which decisions about life and death assert themselves with tragic force (1000–1001). *Life of Pi* invites such considerations of life and death, and its magic realist mode allows it to entertain the possibility that a radical decision can be indefinitely postponed. Genre here functions as a strategy for enchantment, not as a conduit for sensation (as it does in Poe); but like in Poe, it serves to evade rather than confront nonhuman agency.

The last part of the novel brutally undoes that evasion as it admits the inevitability of violence and death, only to end up performing another gesture of avoidance. This last part presents an interview between two insurance officials, working for the Japanese Maritime Department, and the older Pi. These officials refuse to believe Pi's fantastic story, and Pi responds by replacing it with a horrific account without animal actors, in which a shipwrecked Pi witnesses acts of cannibalism and the beheading of his mother, after which he himself kills the French cook who murdered his mother. The implication is clear: the story that has entertained us for almost 400 pages is a displaced version of a deeply traumatic experience, a coping strategy through which Pi survives traumatic loss. What seemed like a fabulous encounter with the nonhuman world is part of a psychologically realist scenario of coping with human violence. The end of the novel, in other words, reinstates the generic conventions of realism, most notably psychological and traumatic realism, to make sense of the strangeness of the preceding novel. This strangeness, featuring weirdly subdued tigers and meat-eating islands, is ultimately thoroughly rehumanized. As in *Arthur Gordon Pym*, the novel ultimately fails to fit the nonhuman world into a generic pattern, and it ends up paradoxically and indirectly pointing to the unsettling reality of the Anthropocene.

Mat Johnson's (2011) comic novel *Pym* spins another affective and tonal variation on Poe's novel. Mainly a satire of contemporary identity politics, *Pym*'s unashamed ludicrousness qualifies it as what Mark McGurl (2012) has called a work of "posthuman comedy": a farcical engagement with human life's diminishing stature in the order of things. Chris Haynes, an unsuccessful black professor of American literature, discovers that Dirk Peters, Arthur Pym's monstrous (and magically whitened) companion in Poe's novel, actually existed. He organizes an expedition to recover the all-black island of Tsalal, which, in the context of contemporary identity politics, is no longer a site of dread (as it was for Poe) but a "great undiscovered African Diasporan homeland, uncorrupted by Whiteness" (39). Except it isn't: Poe's sublime figure of whiteness has spawned a population of white monsters who enslave Haynes's all-black crew to do mining work underneath the permafrost of Antarctica, until Haynes manages to escape to the "Dome of Light," a biodome designed by Thomas Karvel (a transparent satire of American kitsch painter Thomas Kinkade, the self-declared "Painter of Light™") and inhabited by humans rather than white monsters. The novel's irreverent tone, even when dealing with such issues as genocide and extreme violence, underlines its ambition to avoid the clichés of Antarctic literature and debunk the heroism of "yet another polar epic of man succumbing to nature" (94). Neither sublime nor traumatic, *Pym* is appropriately preposterous.

Pym not only remixes the elements of Poe's novel (even Pym is still alive after two hundred years!), including its color-coding, as it also uses its satirical edge to target another Anthropocene fantasy: that of the self-contained and self-sustaining biodome. Biodomes are closed ecological systems that are supposedly independent from the outside world, even though the most famous of such experiments, that involving the so-called "Biosphere 2" in the 1990s, was notoriously unsuccessful. And so it is in Johnson's novel: life in the biodome is accompanied by the continuous drone of the fossil-fuel-driven engines keeping the dome warm and livable. In this way, the novel shows the ecomodernist idea of a "good, or even great" Anthropocene that "decouples" from the outside world to be an illusion—an illusion, moreover, that relies on the continuation of racialized exploitation. Karvel's design of the dome, for which he even painted the sky, also suggests that literature and art might be complicit in rendering uncomfortable realities attractive or invisible. The novel remarks that Karvel's world "seemed a place where black people couldn't even exist, so thorough was its European romanticization" (184). Indeed, the novel's flippant recycling of familiar tropes of Antarctic literature (lost tribes, underground civilizations, purported utopias) demonstrates that Antarctica is less a nonhuman outside to culture than a thoroughly mediated imaginative site. The notion of Antarctica as a blank outside to culture, as a place untainted by human destruction, is congruous with the obsession with white purity that is on display in Karvel's painting and in Poe's novel. As the novel notes, white people preserve their imagined whiteness by "refusing to accept blemish or history. Whiteness isn't about being something, it is about being no thing, nothing, an erasure" (225). The allure of the nonhuman world, the novel suggests, is akin to the desire for historical impunity and innocence.

The novel's own farcical mode is then a way of recognizing the inevitable messiness and leakiness of all ecological constellations, and the impossibility of insulating lifeboats, spaceships, and biodomes. Its climax emblematizes the uneasy overlap between questions of race and climate change. The exhaust of the biodome's engines threatens the life-world of the white monsters (a fear they refer to as "the Melt" [196]), and the novel's protagonist provokes an attack by the monsters (slaveholders as well as climate-change victims!) on the dome to be able to escape. Escape, that is, depends on an act of genocide. If Poe's novel ends on a vision of splendid whiteness, *Pym* ends (equally abruptly) when the protagonist discerns what the last sentence refers to as "a collection of brown people, and this, of course, is a planet on which such are a majority" (322). Brown evades the strict division between white or black, and it points beyond the destructive fantasy of a self-contained, mono-racial community (K. Davis 2017, 42). The novel, for Johnson, is not a microcosm that allows us to imagine a micro-solution to the challenges of the Anthropocene. Instead, it is part of a leaky ecology in which genres refract the entanglements of forms of life and life forms. Compared to *Life of Pi* and *Arthur Gordon Pym, Pym* seems more clear-eyed about the unstable relation between genre and reality. And, we may well ask in this Anthropocene context, what reality? What world?

Earth, World, Globe, Planet

The expedition in which Poe's Arthur Pym participates not only wants to map Antarctica, it also aims to establish the ontological status of the continent: it sets out to "solv[e] the great problem in regard to the Antarctic continent" (Poe 1999, 161), that is, the question whether Antarctica is a substantial, stable continent at all, or just a drifting ice pack (like the Arctic, which was much better known at the time Poe was writing). When Pym has convinced himself that it really is a solid part of the earth, he suddenly feels an earthquake—the calamity he will go on to attribute to the natives—which unsettles this steady ground: it is, he notes, as if "the whole foundations of the solid globe were suddenly rent asunder, and that the day of universal dissolution was at hand" (183). Another way in which *Arthur Gordon Pym* anticipates recent Anthropocene discourses, then, is in its undermining of categories such as globe (and globalization), world, earth, and planet. This is yet another place where literature can contribute to Anthropocene discourses: not only is literature traditionally seen as a form of imaginative world-building, the previous section has also shown that literature can point to the *limits* of imagined worlds (such as, for instance, the lifeboat, the biodome, or the ship) and establish new constellations between globe, world, earth, and planet—all key terms in discussions of the Anthropocene.

Literature's affinity with world-building makes it a promising site to question processes of *globalization*, a category that captures the erosion of nation-states and the increasing economic, cultural, social, and (geo-) political connectedness of all parts of the planet through technology and capitalist markets over the last few decades (Szeman 2005). Globalization imagines the world as a single space and pictures it as an extended grid in which all areas are mapped and measured. The globalized world is a world that can be controlled; it is, in the words of the French philosopher Jean-Luc Nancy (2007), a realm of "global equivalence" and of "indistinct integrality" (54, 27). Like Jacques Derrida, Nancy invokes the distinction (unavailable in English) between *globalisation* and *mondialisation* to argue for a form of "worlding" that is not just reducible to spatial extension and to capitalist calculation (Derrida 2005). Instead, *mondialisation* has a distinct history, recognizes differences, and imagines values that are not reducible to prices. In his book *What is a World?*, literary scholar Pheng Cheah (2016) develops a sophisticated argument that literature's capacity to create worlds empowers it to disturb the smooth and frictionless circulation of capital and to remind audiences of global unevenness and of horizons of meaning beyond capitalist exchange. Literature, for Cheah, shows that the world is never exhaustively mapped and controlled; it is fundamentally open, and "this openness is an unerasable normative resource for disrupting and resisting the calculations of globalization" (9). If capitalist globalization thinks it has saturated and mastered the globe, literature becomes a placeholder for "alternative cartographies" (17). The literary reimagining of Antarctic isles as sites of horror (in Poe), of carnivorous threat (in *Life of Pi*), or of posthuman comedy (in *Pym*), for instance, reminds us that the South Pole is more than a site of resource exploitation and military territorialization; it counters the continent's reduction to the status of a mere part of a globalized but ultimately worldless world.

There is one danger with the celebration of literature's world-building capacities: an insistence on "counterglobalist" worlding risks tipping over into an illusion of transcendence, an illusion that we can abandon the earth for an artificial reality of our own making. We can think here of the biodome that *Pym* satirizes, or of Elon Musk's widely-publicized ambition to make humankind "a truly multi-planetary species" by colonizing other planets (Musk 2017, 46). Hannah Arendt, one of the most brilliant philosophers of human world-building, opens her 1958 masterpiece *The Human Condition* with a reflection on Sputnik, the first artificial earth satellite that had been launched by the USSR the year before. For the first time, people, "when they looked up from the earth toward the skies, could behold there a thing of their own making" (Arendt 1998, 1). This is an achievement of momentous importance: it is, for Arendt, "the first 'step toward escape from men's imprisonment to the earth'" (1). Arendt does not see this as a movement of liberation, but as a massive mistake: such a "fateful repudiation of an Earth who was the Mother of all living creatures," for Arendt, forgets that "[t]he earth is the very quintessence of the human condition" (2). That we share our life on a finite planet with limited space and limited resources is constitutive of the political dimension of human life, as it reminds us that shared deliberation and collective decision-making is as necessary as it is challenging. Promoting the fantasy of a life elsewhere erodes our awareness of the fragility of our collective forms of life.

Arendt's unease about the Sputnik satellite underlines the need to complement the affirmation of "counterglobalist" world-making with a recognition of limitation and boundedness, even if recent reflections on the category of the *earth* (the term under which these issues are often debated) often push that recognition beyond Arendt's resolute humanism. For Kelly Oliver (2015), for instance, the earth names a limit to human worlding: an "earth ethics" accepts "that we are limited creatures who are not just living on earth, but rather part of the biosphere that constitutes its very being" (39). "At bottom," Oliver writes, "the earth is a limit against which totalizing tendencies of technology abut. The earth always and necessarily juts through the globe to remind us of our own limits" (40). Even if we do not all share the same world—as my values, norms, and conceptions may well be very different from yours—we do share an attachment to the earth. As I already mentioned in the introduction, for Bruno Latour (2017), accepting the challenge of the Anthropocene means accepting that we are inevitably "earthbound." As the firm distinction between the social and the natural world has exploded, we find ourselves "on land shared with other often bizarre beings whose requirements are multiform" (38). "The Earthbound" accept limitation, entanglement, and discord, and they know the main threat comes from "Humans" who still believe the planet is a gridded globe to be subdued: "Whereas the Humans had '*Plus ultra*' as their motto, the Earthbound have no motto but '*Plus intra*'" (291).

Postures of earthly attachment and cosmic detachment have in common that they relate to the earth *as a whole*—that is, to the earth as a *planet*. Such a relation was decisively shaped by images of the earth from outer space. In 1968, over a decade after the launch of the Sputnik, the Apollo 8 mission produced an iconic image that sees the earth emerge above the moon's surface—a picture commonly known as *Earthrise*. The 1972 *Blue Marble*, which portrayed the planet as a precious yet precariously isolated entity against a black background, was an even more formidable catalyst of the environmental imagination. These images made the planet visible as something different than an extended gridded surface: it showed an internally varied, dynamic, colorful yet fragile habitat, and showed that "Earth is the only body that looks even remotely alive from that vantage point" (Oliver 2015, 15). In a sense, then, *Earthrise* and *Blue Marble* "'reterrestrialized' the globe. They turned the globe back into Earth" (Lazier 2011, 623). They inspired an image of the planet as an interconnected, self-regulating system that sustains the conditions for life, as in James Lovelock and Lynn Margulis's popular Gaia theory, or as in Buckminster Fuller's image of Spaceship Earth (already launched in 1963), which figures "the Sun, the Earth and the Moon [as] nothing else than a most fantastically well-designed and space-programed team of vehicles" (Fuller 2001, 290–91) and sees the earth as an integrated machine, "a technological wonder that unites mankind" (Oliver 2015, 15).

The visual cue of a single planet does not automatically lead to an awareness of earthly limitations: it can also sustain the ecomodernism *avant la lettre* of someone like Fuller. Lovelock and Margulis's conception of the earth as an interconnected organism is not that different from Fuller's notion of the earth as a smooth-sailing spaceship: for Lovelock, too, Gaia is "the totality constituting a feedback or cybernetic system which seeks an optimal physical and chemical environment for life on this planet" (Lovelock 2000, 10). The notion of Gaia cannot be understood apart from the ambitions of cybernetics, which aimed to mobilize computational thinking to erase the distinction between communication systems and living organisms. The cybernetic ambition to control and program human behavior emerged in the militarized context of the Cold War, and so, not coincidentally, did the space program that produced the *Earthrise* and *Blue Marble* images. If these planetary images promise cosmopolitan connectedness, they also rely on technology that cannot be divorced from military contexts. One aspect of *Blue Marble* that makes the earth look fragile and vulnerable is the visual mastery conveyed by the scale and detail of the picture: such mastery, we know, is generally an accomplice of military powers of destruction. In a classic essay, anthropologist Tim Ingold (2000) distinguishes between the imaginaries of a "sphere," which designates an environment we live in, and that of a "globe," which evokes an external surface "*waiting to be occupied*, to be colonized" (214). Globes, on this account, are always viewed from outside, and are abstracted from lived experience. Matthew Taylor has shown how the very idea of planetarity emerged in the early twentieth century in the context of imperialism and war. At this historical juncture, Taylor (2016) writes, "the paired revelations of the cosmos's enormity and the earth's shrinking frontiers coalesced to produce a humanly circumscribed globe" (115). These contexts make it uncertain whether fashionable discourse about the planetary can surrender these violent overtones and leave room for diverse kinds of lived experience.

The relation between, world, globe, earth, and planet has become very intricate: world-building, we saw, can serve as a counter to globalization; the earth can offer a limit to unrestrained world-making; and the planet(ary) functions *both* as the condition of possibility of such terrestrial imaginaries *and* as an occasion for subduing and controlling the earth; and this desire for control brings us right back to the fateful reductions of globalization. Concerns over the destabilization of the very earth we inhabit threaten to turn into a vicious loop, in which repeated interruptions, whether they occur in the name of the world, the earth, or the planet, do not keep the world from spinning out of control. The difficulty of interrupting this cycle has resulted in a tendency in critical thought to insist that a particular part of the planet radically resists reduction and colonization. This site of resistance is *not* a particular location. Instead, critical thought points to an irreducibly alien dimension that pervades our everyday lives, and hides in the cracks and fissures of the worlds we construct. Bill McKibben's coinage *Eaarth*, as a new name for an irrevocably altered planet, is probably the most clumsy such effort, even though the other possible names McKibben (2010) suggests—"Monnde, or Tierrre, Errde, ёккучивать"—suggests things could have been worse (2–3). Frédéric Neyrat (2017) talks about an "eccentric" or an "unconstructed" earth to name the dark side of the planet that the *Blue Marble* image cannot capture. The eccentric earth is the planet's "nocturnal, unobjective, and asubjective part"; it is "the earth whose being eternally eludes its spherical aspect" (8–9). Eugene Thacker's work shares this concern with the gloomy and the horrific: it calls this elusive dimension "the world-without-us" or simply "the Planet" (as opposed to "the World," which is "the world-for-us," and "the Earth," which is "the world-in-itself" [2011]). Thacker's word choice joins other critical voices, such as Jennifer Wenzel and Gayatri Spivak, who use the notion of the planetary to name forces that destabilize "the hegemony of the global" and to refer to an inevitable remainder that disturbs the composure of all too human ways of apprehending the world (Wenzel 2014, 21). This world-without-us, for Thacker (2011) and others, is not a particular region: it is simply the world insofar as it is not given to us that intrudes "in the very fissures, lapses, or lacunae in the World and the Earth" (8).

As we have already seen, Thacker promotes science fiction and supernatural horror as genres that can provide access to this shadowy side of the Anthropocene world. Complementing Thacker's case for genre fiction, I have suggested that the self-interrupting dynamic of literary world-making makes it possible to adumbrate the world-without-us in the gaps and cracks between generic and formal templates. This does not mean that literature is the only medium that affords such experiences. What it *does* mean is that it is rewarding to consider the world-without-us as more than a *thematic* concern and as essentially connected to the ways specific genres, forms, and media engage reality. In the next section, I show how different media mobilize their specific medial features to explore the limits of their world-making. I focus on photography, cinema, and post-cinema, before I conclude with two examples of literary works that evoke nonhuman dimensions by drawing on the very features that define them as works of genre fiction or serious realism. Like the non-literary examples, they end up turning the world we thought we knew inside out.

Media Affordances: Photography, (Post-)Cinema, Literature

The contemporary environmental imagination abounds with figures of earths, worlds, and planets: think of post-apocalyptic imaginings of *worlds without humans* (which I discuss in my last chapter): of fictions of *global hyperconnectedness* making the planet vulnerable to instant pandemics, which is routinely reflected in images of civilizational breakdown and zombie apocalypse; or of works like N.K. Jemisin's multiple Hugo Award-winning *Broken Earth* trilogy, which reimagines humans' weirdly intimate relations to the earth through a group of people who have the power to start earthquakes or lift mountains. Still, and as I have shown earlier in the chapter, the relation between cultural artefacts and Anthropocene reality is not only a matter of the imagination: literature and other media can also indicate the unsettling realities of the Anthropocene by exploring the limits of their world-building capacities. Importantly, these capacities are *medium-specific*; literary world-building is different from, say, the world-building in video games. Analyses of cultural works' engagement with the Anthropocene, then, must take the specificity of particular media, genre, and forms into account.

Jennifer Fay's 2018 book *Inhospitable World* argues for the deep affinities between the medium of *cinema* and the Anthropocene. Fay situates the privilege of cinema in the basic fact that, in order to make a film, sets need to be built, weather-events need to be simulated, natural settings need to be meddled with … Cinema affects and even constructs the environment in a far more literal way than literature, and this, for Fay, makes it powerfully analogous to the Anthropocene's transformation of the world into an "unhomely" place. Fay gives the example of some of Buster Keaton's films from the 1920s, which achieve their comic effect partly through calamitous weather that is clearly manufactured and does not hide its artificiality, and in that way foreground "anthropogenic environmental change and modern—at times tragic—modes of inhospitable world-making" (2018, 28). Literature, in contrast, "has nothing of the materiality of the film studio or the temporality of film production; nor does it share with cinema the ambitions of environmental design" (8). Literature, for Fay, lacks the concrete and fragmentary dimension of cinema, and always constructs an achieved totality in the reader's imagination.

Fay's case for the intimate connection between cinema and the Anthropocene is convincing, but it underestimates literature's self-interrupting power: its capacity to dismantle an imagined world even while it is constructing it. To be sure, the production of literature lacks the recalcitrant material reality that cinema does encounter in the production process; Fay is right that the observation that "the production of books disappears trees" (8) makes a fairly weak case for the environmental imprint of literature. Still, literature's self-impeding operations do unsettle the relations between world, earth, and planet in ways that resonate with the Anthropocene's reorganization of reality. If cinema engages with the material world in a more *literal* way, interpreting that engagement as a way of dealing with the Anthropocene, as Fay does, still depends on an *analogy*, and thus on *figurative* thinking. Fay formulates this underlying analogy as follows: "The Anthropocene is to natural science what cinema … is to human culture. It makes the familiar world strange to us by … transforming and temporally transporting humans and the natural world into an unhomely image" (3). Cinema and the Anthropocene encounter each other in a web of meanings that is not merely literal.

Cinema "mak[ing] the familiar world strange" recalls Viktor Shklovsky's case for literature's defamiliarizing force that I touched on in the last section of the previous chapter. Cinema is then not distinctive because it, in Fay's words, "helps us to see and experience the Anthropocene as an aesthetic practice" (4), something other media can also do, but because it has particular resources at its disposal that literature and other media lack. These medium-specific affordances are especially clear in works that intimate the "unhomely," "inhospitable" world of the Anthropocene without being thematically invested in environmental issues. Fay gives the example of 1950s and 60s film noir, a genre that is not thematically concerned with planetary destruction, but which is situated in gritty "urban locales about to be demolished" and features doomed characters addicted to the kind of "bad living and unsustainable striving that underwrite the accumulating culture of the Anthropocene at midcentury" (18). Film noir's decaying infrastructures and melodramatic narratives are, moreover, deeply connected to the affordances of celluloid, a medium that lends itself particularly well to a saturated black and white (rather than midtones), but also a form of plastic that is imperfect, organic, and bound to decay. Factoring in this medial dimension makes it possible to read film noir as "a kind of extinction narrative" (18): film noir tells the story of how the human species has rendered its habitat inhospitable, even while this habitat will outlive our death-bound species.

Joanna Zylinska has argued for a particular affinity between *photography* and the Anthropocene. That rapport may be evident in the work of someone like the Canadian photographer Edward Burtynsky, for instance, whose large format images of devastated natural sites and manufactured (but mostly depopulated) landscapes has, like the equally majestic urban and (post-)natural landscapes of Andreas Gursky, deeply shaped the visual signature of the Anthropocene. For Zylinska, the connection goes beyond such sublime powers of representation, and is anchored in photography's status as a geological process. Geological processes are defined by their capacity to leave an imprint on the planet, which then becomes legible as a fossil; photography, in its turn, is a process in which reality imprints a photochemical surface that becomes visible as a photograph. This photographic process thus offers a way to redescribe the way reality leaves a trace: the past, Zylinska (2018) writes, literally "photographs itself" (54). Zylinska insists that "the link between fossilization and photography" is more than a metaphor, and that lingering on this connection makes it possible to appreciate that the conditions under which photography can exist are the same under which human life can survive: light, energy, and the sun (54). If photography and fossilization are similar processes, photography becomes a medium dedicated to intimations of extinction, of a planet where human life will have become a mere fossil. For Zylinska, this ominous suggestion is even—or especially!—there in schmaltzy, romantic images of sunset pictures. This power to reflect, or this *in*ability *not* to reflect, on species extinction is *not* a consequence of thematic choices, but a feature of the very medium of photography: because of its famous "indexicality," its unavoidable reliance on physical reality that sets it apart from painting, drawing, or indeed literature, photographs cannot avoid being the vulnerable imprint of a finite reality.

Of course, the relations between cinema, photography, and extinction are altered once we recognize that visual culture is now overwhelmingly digital. Altered, but not mitigated: the vast computational powers that generate contemporary images that, in their turn, overload our human information-processing capacities offer us a glimpse into a reality that does not require a human viewer. The term *post-cinema* underlines that the contemporary media landscape no longer relies on the twentieth-century sensibilities reflected in and shaped by cinema and television (Denson and Leyda 2016). For Shane Denson (2018), post-cinema is not only thematically obsessed with extermination events (whether in the slowed- and pared-down bleakness of a film like *The Road* or the uncontrollable proliferation and acceleration of zombie bodies in *World War Z*), its microtechnical basis makes possible "a general *discorrelation* of moving images from the norms of human perception and embodiment that governed classical cinema" (1). If classical film provided viewers with cues about where to look and directed their perspectives and emotional responses, post-cinema often offers "too much visual information, presented too fast for [viewers] to take in and process cognitively" (8). It has the power to overload images in a way that only allows for fleeting, dispersed scanning. Denson gives the example of the "hyperinformatic" action scenes in films like the *Transformers* franchise, which revel in computer-generated images that provide an excess of visual information in too short a time for the human eye and brain to process. Even if these are hardly films with a considered environmental message, they engage extinction on a fundamental level: such films leave viewers with a sense that the products of post-cinema's technological infrastructures are strictly indifferent to their perceptual attention; they are not interested in figuring a "world-for-us," but offer a glimpse into a "world-without-us."

If Eugene Thacker, as we saw in the previous section, found such intimations in the "fissures, lapses, or lacunae in the World and the Earth" (2011, 8), post-cinema generates them through the very *seamlessness* of its simulations: through its "discorrelation" from merely human perceptual and cognitive capacities, this incredible seamlessness annuls the terms of the "loose affectual contract" that, according to Lauren Berlant, organizes the relation between a cultural object (in this case, traditional cinema) and its audience (2008b, 847). Yet crucially, such a destabilization of emotive expectations is itself an affective event, even if the affect that is generated is rarely cathartic and euphoric, and more often disparate, awkward, and uncomfortable. Denson writes that post-cinema "envisions and transmits affective clues about a world without us" (2018, 1); in a comparable way, we have seen how the Antarctic formations of Poe, *Life of Pi*, and *Pym* produce disorienting affects through the gaps *between* the different genres they enlist, and not only *within* the terms of the affectual contracts that sustain these different genres' world-building operations. What these examples have in common is that they all capitalize on genres' and media's capacity to disorganize the expectations that audiences bring to them. They have the power to interrupt their customary operations and hint at realities that remain foreign and render the world they depict inhospitable.

I began this chapter with the tenuous distinction between "ludicrous" genre fiction and "serious" realism; I want to end with two examples that demonstrate that both forms are connected by their capacity to disturb their own operations and indirectly hint at a less homely reality. My first example is a work of genre fiction. N.K. Jemisin's wildly successful *Broken Earth* trilogy (the winner of three consecutive Hugo Awards for Best Novel) begins with an arresting sentence: "Let's start with the end of the world, why don't we?" (Jemisin 2016, 1). The rest of the trilogy refuses to set this question aside as a merely rhetorical one. Its narrators entertain the possibility that it might be preferable to end the world they inhabit. The story is situated in a future continent ironically called "the Stillness," in which cataclysmic geological events constantly threaten the different "comms," and in which misogyny, violence, exploitation, and precarity are rampant. Jemisin's decision to people her story-world with characters who are trans, queer, and people of color shows how her books break with the template of the typical science fiction story, in which, as Jemisin explains in an interview, a white male hero uses "the opportunity of the apocalypse" to act out "white male power fantasies" (Hurley 2018, 469). In the *Broken Earth* trilogy, apocalypse is not an opportunity for escape from the status quo, as the status quo *is* a prolonged apocalypse marked by intermittent acts of gendered and racist violence against the books' vulnerable protagonists and by the imminent threat of climate destruction.

By frustrating the generic expectations of the SF genre, the trilogy unsettles the reader's affective attachments to the world that the genre commonly reinforces. The novels' emotive work generates a more troubling affect: the sense that starting with the end of *this* world might not be the worst idea; that, just perhaps, a world like this doesn't deserve to be saved. The world whose undoing the trilogy contemplates is a version of the Anthropocene present. This is underlined by the particular nature of the vilified minority on which the books focus: orogones are people with the ability to contain and provoke geological catastrophes through telekinetic (but deeply embodied) powers. The orogones are necessary for the survival of life in the Stillness, but they are also shunned, threatened, and killed because they are dangerous and different. The orogones are not only allegories for the reviled but necessary adjuncts of the capitalist world system—women, people of color, exploited workers; their intimate connection to the forces of "Evil Earth" also signals the novels' awareness that socioeconomic challenges (to do with *forms of life*) are fundamentally entwined with questions of planetary belonging (implicating the human *life form*). It is the combination of the *Broken Earth* books' generic experiments and their socioeconomic sophistication, then, that evokes a sense of unease with our current constellation of world, earth, and planet.

Few recent novels display their high cultural credentials as ostentatiously as Teju Cole's 2011 debut *Open City*, which will be my last example. Narrated from the perspective of the artistic and highly educated Nigerian-American Julius as he walks the streets and travels on the public transport systems of New York and Brussels, the novel is ostensibly a celebration of the power of literature, high culture, and art to foster intercultural curiosity and cosmopolitan connectedness. *Open City*'s virtuoso entanglements of masterpieces from different geographical and historical locations seem to illustrate how the levelling forces of *globalization* can be countered by the sublime *world* of art and culture. The novel's psychological realism seems perfectly suited for such world-building: the first-person perspective grants the reader insight into the texture of a refined and flexible mind processing impressions and memories, and thus shows cosmopolitan world-building in action. Yet this is not quite what the novel delivers: instead, it turns these expectations inside out by showing readers how Julius, for all his sophistication and apparent curiosity, remains strangely isolated and unchanged by the stories and encounters that are supposed to enrich him. The fragments of cultural traditions never add up to a new world. And as in many of the other works we surveyed in this chapter, this failure opens cracks and clefts that hint at a less homely world, at a less hospitable planet.

Open City uses the affordances of psychological realism to show that Julius's stream of consciousness contains a more ominous dimension.

The plot makes this explicit when, near the end of the novel, it reveals that this paragon of cultivation and memory has apparently forgotten that he raped a friend's sister. Another episode near the novel's end connects this sinister undertone, which spoils the refined feelings that cosmopolitan fictions are means to generate, to a sense of planetary derangement. Julius finds himself locked out of a Mahler performance at Carnegie Hall (Mahler is the composer, the novel duly notes, of *Das Lied von der Erde*). Standing on "a flimsy fire escape," Julius describes his situation as one of "unimprovable comedy" (255). This comedy doesn't improve, but it does change into something close to McGurl's "posthuman comedy" when Julius beholds the stars, as he is high enough not be surrounded by light pollution, and reflects that "their true nature was the persisting visual echo of something that was already in the past" (256). Julius understands that many of the lights he sees were emitted by light sources that have by now been long extinguished, while in the darkness, light might already be underway that has simply not yet reached the earth, and might do so only "long after the human race itself was extinguished" (256). Some stars are dead and shining, others are alive and dark. The novel here makes explicit what was already implied by its formal manipulation of the conventions of psychological realism: that life on earth is, strictly speaking, already dead. Literary language, as I have argued, is not the only medium that can make us feel that, but like other media, it makes us feel it in medium-specific ways.

3 Objects, Matters, Things

Other-than-human(ism)

The Anthropocene launches an ambivalent challenge to the belief that our species holds special powers that set it apart from other life forms. If naming a geological epoch after our species elevates us to the status of a veritable geological agent, it does so only to underline our impotence and vulnerability in the face of the forces we have helped unleash. The Anthropocene renames the recent past as an interval of momentous human agency, but it also dramatically shortens the historical window in which that agency can still undo the unintended consequences of its actions. A 2018 report by the United Nation's Intergovernmental Panel on Climate Change, for instance, situates an all-out climate disaster as early as 2040, with the deadline for intervention expiring much earlier still. If in the Anthropocene, "the human along with every other being is intimately caught within a maelstrom of erosion and disintegration" (Oppermann 2018, 3), the traditional markers of human species pride—self-consciousness, autonomy, reason—merely point to an exceptional capacity to register impotence and doom, not to mention an intensified species status anxiety.

The increasing entanglement of human and other agents and the growing awareness of vulnerability inform one of the most important developments in contemporary critical thought: a decentering of human life and a growing concern for nonhuman lives and things. The so-called "nonhuman turn" (Grusin 2015) downplays the differences between humans and nonhumans, and often situates different human and nonhuman agents as participants in a "flat" ontology: a worldview without hierarchies and layers, in which agency is distributed between interacting human and nonhuman actors. As we will see, this does not simply mean that humans become more thing-like. More often, it means that objects are endowed with humanoid capacities: the power to withdraw into an inscrutable interiority (in object-oriented ontology), to express themselves (in the new materialism), or even to tell stories (in material ecocriticism). These paradigms officially claim to move *beyond* the humanist belief in human singularity, yet they often turn out to be "*ultra-humanisms*" (Colebrook 2014, 162), in that they extend human attributes to nonhuman entities. Even if these capacities are mostly metaphorical, this transfer testifies to the complex traffic between human and nonhuman forces in contemporary theory.

In this chapter, I survey the contributions of object-oriented ontology, new materialism, material ecocriticism, and (much more briefly) actor–network theory and critical posthumanism to debates over the relations between human and nonhuman agency. My discussion shows that these theories take great pains to appreciate difference and not surrender the human and nonhuman things they bring together to a vast, undifferentiated sludge. Their attention to discrete things and separate objects, however, does not always translate into an ability to differentiate between different kinds of differences—an ability that, as we saw in the first chapter, characterizes the environmental humanities at its best. Take, as a famous example, political theorist Jane Bennett's analysis of the 2003 North American power blackout in her book *Vibrant Matter*. The breakdown of the electric power grid, which affected 50 million people, helps Bennett (2010) illustrate her notion of "*distributive* agency" in which the cooperation and "interactive interference of many bodies and forces" has a real-world impact surpassing that of any individual agent (21). Each of these constituents is given their due in Bennett's catalogue: "the electrical grid is better understood," she writes, "as a volatile mix of coal, sweat, electromagnetic fields, computer programs, electron streams, profit motives, heat, lifestyles, nuclear fuel, plastic, fantasies of mastery, static, legislation, water, economic theory, wire, and wood" (25). The blackout, in Bennett's analysis, is then the "end-point of a cascade" (25): it is triggered by the frictions and overlaps between these constituent parts, which serve as different "agential loci" (26) contributing to the agency of the assemblage they compose.

Bennett refuses to discriminate between the different components. Even when we know that corporate greed and "the shabby condition of the public-utilities infrastructure" (36) play a role, Bennett declines to differentiate between, say, the power of corporations and, say, the role of coal, sweat, or electronic transmitters: these things are different, but they are all different in the same way. This emphasis on the agency of things (Bennett talks about "thing-power") is a welcome corrective to the almost exclusive focus on social forces and linguistic mediation in the late-twentieth-century humanities. Against such a stale "constructivism," more recent theoretical tendencies affirm what Rosi Braidotti (2013) has called the "vital, self-organizing and yet non-naturalistic structure of living matter itself" (2)—and in the case of Bennett's vibrant materialism, even of abiotic matter. Recent critical projects aim to undo the so-called "linguistic turn," which in the last third of the twentieth century shifted attention to the ways language shapes and polices access to reality. Instead, they present reality as a choreography of human and nonhuman agents: "Humanity and nonhumanity have always performed an intricate dance with each other," Bennett writes (31). Such a consideration of the material world was long overdue, not least because an exclusive focus on language threatened to obscure the realities of climate change and other processes of planetary deterioration. Still, Bennett's principled refusal to single out anthropogenic dynamics—to differentiate between *different differences*—also has less salutary consequences: it downplays the disproportionate responsibility of particular human agents in environmental degradation, and it elides power differentials between diverse human constituencies. In their ambition to correct the constructivist bias of the late-twentieth-century humanities, projects like Bennett's at times derail the environmental humanities' mission to negotiate the entanglements *and* differences between symbolic and material processes.

In the next section, I trace continuities and differences between the most prominent contemporary paradigms that attend to other-than-human forces, most notably the new materialism and object-oriented ontology. The third section of the chapter develops a brief reading of Scottish writer John Burnside's 2008 novel *Glister* to test the limits and affordances of these theories. Set in a toxic postindustrial landscape, *Glister* maps the intricate connections between technological, natural, and human agencies. The novel resonates with these critical paradigms' attention to the "thing-power" of the environment, xenobiotic substances, and industry, and their insistence that human lives are co-constituted by nonhuman substances. At the same time, it shows that these theories have a hard time calibrating differentiated human responsibilities and issues of class. The fourth section restates the failure of these critical approaches to factor in hierarchies and incompatible differences as a matter of scale. Scale is at the heart of the notion of the Anthropocene, if only because the very term suggests a smooth scaling up of human agency to the level of the species. I draw on Timothy Clark's notion of "scale effects" and Derek Woods's term "scale variance" to underline that size matters: strategies that work on a personal or local level (such as recycling and ethical consuming, for instance) cannot simply be expanded to a planetary scale. In the fifth and final section, I discuss the material ecocriticism of Serenella Iovino and Serpil Oppermann. For these critics, all matter is "storied matter": it has the power to be "creatively expressive" and tell its own stories (Oppermann 2018, 9). I end with the suggestion that material storytelling might be a less productive literary model for thinking about the imbrication of human and nonhuman forces than writing. Writing and inscription may be more adequate models to convey how violent, consequential, and inerasable human interactions with the planet have been and continue to be.

The theoretical developments I discuss break with a humanist investment in the exceptionality of human life, and they explicitly attend to other-than-human agents. This brings them close to *posthumanism*, even if, as I noted, these theoretical movements remain hooked to human capacities (storytelling, intentionality, expression). But critical posthumanism is also better understood as a move beyond human*ism* than as a relinquishing of everything human. For Rosi Braidotti, posthumanism consists in a decentering of the traditional autonomous modern subject, who represents only one version of human life: what she calls "the human of Humanism," who forecloses difference and "stands for normality, normalcy and normativity" (2013, 26). For Cary Wolfe (2010), also, posthumanism is first of all posthuman*ist*: it "opposes … fantasies of disembodiment and autonomy" and emphasizes human life's "imbrication in technical, medical, informatics, and economic networks" (xv). In the wake of Donna Haraway's (1991) influential "Cyborg Manifesto," posthumanist thought has foregrounded technical and robotic elements, and its emphasis on complex environmental interactions and the basic openness of systems has made it a privileged interlocutor for the environmental study of literature. As we saw in the first chapter, ecocriticism has increasingly come to expand its scope from the natural wilderness to environments marked by toxicity, invasive species, intelligent machines, and pesticides. In this context, critical posthumanist thought enriches the ecocritical toolbox for analyzing the complexly interwoven biospheres and technospheres of the Anthropocene. Like critical posthumanism, the theories under scrutiny in this chapter help Anthropocene thought move beyond anthropocentrism and map the complex material and symbolic traffic between humans and nonhumans.

Matter vs. Object

One way to describe the *new materialism* is as a project to *make matter matter*. Matter, for the new materialism, is not a passive substance that science can explain and technology can manipulate; nor is it merely meaningless goo that only acquires meaning through linguistic mediation. New materialism is first of all a *monism* that undoes the dualisms inherent in the ways we traditionally relate to matter: that between scientific object and observer, for instance, or that between matter and meaning. This does not mean that materialism is anti-scientific: new materialists like sociologist Vicki Kirby and science studies scholar and theoretical physicist Karen Barad, for instance, explicitly enlist quantum physics to theorize material reality as a fluid and open-ended process rather than a fungible resource. Quantum physics teaches that scientific perception is an effect of particular scientific procedures (this means, for example, that the position and the momentum of particles cannot possibly be measured at the same time). New materialists extend this insight and see the material world as a reality that emerges through the interactions of different acts, events, and forces: matter and meaning, the material order and symbolic processes, are fundamentally entangled, an idea reflected in Donna Haraway's (2003) notion of "natureculture" (1).

As humans, we are, as Barad (2007) writes, "a part of that nature we seek to understand" (67), even as our meaning-making practices also co-constitute the natural world. Nothing escapes this intercourse, as the new materialist world leaves no room for transcendence and is fully *immanent*. The new materialist worldview is also fundamentally *relational*: the constituents of reality do not preexist their relations to one another, but only emerge through their reciprocal and multidirectional exchanges. To underline this process of co-emergence, in which there are no discrete objects with inherent characteristics but only entangled agencies that mutually constitute one another, Barad coins the notion of "intra-action" (33). She contrasts this term to "interaction," a word that seems to presuppose preexisting entities, just as, for instance, the once fashionable notion of hybridity does. For the new materialism, matter is neither stable nor unchangeable; it is a net of agencies that human life is plugged into.

Monism, immanence, relationality: these notions also characterize Jane Bennett's *vibrant materialism*, whose concerns overlap with those of the new materialism, even if it is somewhat less interested in science. Like Barad and Kirby, Bennett (2010) holds that matter is not just "passive stuff," "raw, brute, or inert" (vii), as there is "a liveliness intrinsic to the materiality of the thing" (xvi). Bennett prefers the term "thing" to that of "object," which for her is too reminiscent of the subject-object dyad that the new materialism proscribes: things, she quotes W.J.T. Mitchell, signal "the moment when the object becomes the Other" (2). Things and assemblages (irreducibly complex compounds of things) have "an active, earthly, not-quite-human capaciousness" (3) with which human life is co-emergent. Bennett gives the example of Omega-3, a fatty acid prevalent in fish that has a salutary effect on people's mental states. Bennett proposes we factor in the agency of this fatty acid when we describe phenomena (or "assemblages") such as "American consumption" and the "crisis of obesity" (39). We need to recognize, Bennett writes, "a productive power intrinsic to foodstuff, which enables edible matter to coarsen or refine the imagination or render a disposition more or less liable to ressentiment, depression, hyperactivity, dull-wittedness, or violence" (49). On this account, emotive states do not originate in the ineffable interiority of the human mind, but emerge through the interaction of human and nonhuman bodies (an insight we already encountered in Evelyn Reilly's *Styrofoam*, which I discussed in the first chapter).

Picturing the world as a mesh of promiscuously interrelated oozing bodies no doubt has its attractions, but it also has its problems. New and vibrant materialisms tend to celebrate interconnectedness as a nourishing and enchanting state of affairs, but delight and pleasure are only some of the possible moods besetting material encounters: if the material bodies one encounters happen to be toxic or carcinogenic, for instance, danger and harm might replace generosity and joy (Lemke 2018, 40). Another problem is that this world picture brackets epistemological questions (how can we get to know other entities?) only to transform its ontological description of the world into an *ethical* injunction: materialist accounts of the world call on us to "be attentive" to matter, to "attune ourselves" to its energies, or to "register" materiality. With epistemological complications out of the way, all that remains is "the ethical binary of attunement to or resentment to materiality" (Rekret 2016, 227). More often than not, this ethics is an individualized and voluntarist one, in which purportedly free individuals are called upon to cultivate a transformed relationship to matter (Lemke 2018, 42–46; Rekret 2016, 227–30).

At least two politically salient issues remain unexplored in this reduction of human action to a matter of ethical sensibilities. First, and as we will explore in the fourth section of this chapter, such a call for ethical attunement remains blind to the question of scale: real-world change requires more than a cumulative series of individual conversion experiences, and also depends on the hard work of organization and mobilization. It requires a collective politics rather than a personal ethics. Second, the cherishing of autonomy and free will, which the new materialism also extends to nonhuman things, overlooks that human actions are necessarily constrained, and that attitudinal change always takes place in an already stratified social space. Valorizing the agency of Omega-3 is one thing, but the transition to more sustainable food patterns requires more than an altered ethical sensibility: it might, for instance, require undoing material constraints such as the deregulation of the food industry and the socioeconomic determinants of the obesity crisis. Bennett's (2010) proposal to "enter into the proximity of assemblages whose conglomerate effectivity tends toward the enactment of nobler ends" (43) enjoins us to cultivate healthier affects, but "enter[ing] into the proximity" of the good stuff is unlikely to constitute the kind of decisive action that will effectively address Anthropocene challenges.

The new materialism shares its emphasis on relationality and on the choreographies of "horizontally aligned agentic entities" (Oppermann 2018, 3) with *actor–network theory* (or ANT). In the 1980s, ANT emerged in the sociology of science and technology (SST) in the work of Michel Callon and Bruno Latour. It studies the horizontal interactions between social and natural and between human and nonhuman actants. The effect of ANT's descriptions is one of defamiliarization, as infrastructures and everyday objects of lab life—scraps of paper, instruments, coffee cups—are assigned a role in the development of scientific and technological breakthroughs. ANT reimagines the world as a networks of relations and nothing else: "Literally there is nothing but networks" (Latour 1996, 370). It is rigorously uninterested in substances: things only feature in networks as actants, and their identity is exhaustively defined by the mode of their participation in the network under investigation. For Latour, an actant is simply "something that acts or to which activity is granted by others" (373).

The new materialism adopts the relationality and flat ontology of ANT, but it goes beyond it by putting flesh on the bare actantial nodes in ANT networks and by transforming its abstract universe into a dense, material, and embodied one. For both the new materialism and ANT, then, everything that exists is in relation, and reality is the sum of all relations. They share what Peter Gratton (2014) has called "an 'actualism' that grants reality only to the shifting relations of the world and not to hidden forces (even potentially) that don't relate to the things of existence" (91). For ANT and the new materialism alike, things cannot disengage from the networks that sustain them. This marks their main difference from *object-oriented ontology* (or OOO, or Triple O), another wildly influential contemporary school of thought that campaigns for the capacity of things, and not only humans, to withdraw from relation and to preserve a residue of interiority in a hyperconnected world. If the new materialism tends to a cheerful affirmation of the actuality of the world, object-oriented ontology revels in gothic sensibilities and in the horror of the abyss beneath things. The new materialism cherishes "lively things" (Bennett 2010, viii); OOO cultivates the weird, the lurid, and the alien.

Mostly associated with the work of philosopher Graham Harman and literary critic Timothy Morton, OOO emerged in the early 2000s as part of a loose philosophical movement called *speculative realism*. OOO extends the anti-constructivist thrust of a lot of twenty-first-century thought to a wide-ranging critique of the thought of Immanuel Kant. Kant's critical philosophy famously restricts knowledge to the domain of things that appear to us, and it dismisses all attempts to get beyond experience as empty speculation. Speculative realists refer to this epistemological stricture as "correlationism" and reject it to gain access to the reality of things *beyond* human experience. This deeper reality of things cannot be reduced to their molecular constitution (an error Harman calls "undermining") nor to their relations and actions (a form of "overmining" of which ANT is guilty). According to OOO, the new materialism commits both errors simultaneously: not only does it plunge things into a material substrate, it also entangles them in webs of relation. Against this double mistake, Harman (2016)—an indefatigable booster of the OOO brand—positions object-oriented ontology as "a resolutely anti-materialist theory" (95–96): a "deeply *non-relational* conception of the reality of things," he writes, "is the heart of object-oriented philosophy" (Harman 2012, 187). Against materialist flux, object-oriented ontology asserts the unshakable stability of things; against the reduction of things to their relations, it affirms their "irreducible strangeness and surprising weirdness" (Lemke 2017, 144). Things never fully surrender themselves in their encounters with humans, nor in their encounters with one another: "the inner aspect of the object … is forever withdrawn from the sensuous domain" (Gratton 2014, 100).

The affirmation of an ineradicable darkness at the heart of nature obviously resonates with Anthropocene sensibilities. The broad shift from fantasies of natural harmony to a disconcertingly weird environment is reflected in Timothy Morton's blockbuster concept of the "hyperobject." If all objects, for OOO, are strange, wayward, and weird, hyperobjects like plutonium, global warming, Styrofoam, or capitalism radicalize that alienation: hyperobjects are "objects massively distributed in time and space that make us redefine what an object is" (Morton 2011, 167). OOO here reveals its proximity to the figure of the sublime: finding itself overwhelmed and disoriented, it turns away from empirical observation and rational thought to the domain of the aesthetic to find access to things. The writing of OOO-thinkers is then often a self-conscious performance of approximating the object world. The two most remarkable stylistic features of OOO-writing are "ultra-vivid description" (Morton 2011, 170) and the litany: long lists of objects and things that are juxtaposed at a remove from human mediation. One random but entirely representative example is Graham Harman's (2005) evocation of a place "amidst coral reefs, sorghum fields, paragliders, ant colonies, binary stars, sea voyages, Asian swindlers, and desolate temples" (3). Indeed, if the writing of new materialists typically invests in syntactical complexity to express the intricate relations between human and nonhuman objects, OOO relies on a melancholic evocation of the object-world from which human life has ostensibly withdrawn. As Morton writes, "[m]elancholia is precisely a mode of intimacy with strange objects that can't be digested by the subject" (175).

Such seemingly impassive and deadpan evocations of the object-world are not without their problems. For one thing, the stable splendor of objects obfuscates the contexts and contingencies through which the stability of the status quo emerges. The privileging of *stability* makes it hard to account for change and difference, and some of the most scholastic elaborations of the theory follow from the difficulty it has to conceive of change (Gratton 2014, 85–107). This problem is compounded by the simultaneous emphasis on the *inaccessibility* of things: if the only mode of access to "the thing in its untamed, subterranean reality" is aesthetic (Harman 2011, 80), it is hard to see how this improves on the new materialism's recourse to ethical attitudes. Reducing analysis to an exercise in aesthetics, OOO remains blind to "the de facto privileged role and the planetary power of humans to affect other objects" (Lemke 2017, 147). Indeed, one of the ironies of OOO might be that, in its desire to move beyond the "correlationist" confines of human experience, it forgets to problematize the traditional human subject: objects' alleged potential to withdraw from relation and to preserve their privacy is curiously anthropomorphic, and the display of a world in which objects are liberated from their entanglement with humans can only be enjoyed from a perspective that seems to be that of a disengaged and disembodied human spectator. It is no surprise, then, that literary engagements with the object-world diverge from object-oriented ontologies, as well as from new materialist ones. John Burnside's novel *Glister* is a case in point.

Unsafe Spaces, or, The Matter of the Glister

A simultaneous focus on the ineffability and the actuality of nature characterizes the work of the Scottish poet and novelist John Burnside (D. James 2012, 602). It is especially Burnside's 2008 eco-gothic *Glister* that resonates with contemporary Anthropocene discourses, even if, as we will see, it also points to the limits of the theoretical focus on modes of entanglement and attunement. The novel is set in Innertown, a site of social, industrial, and environmental decay situated in the shadow of a disused chemical plant. Toxic spillage from the plant affects all life forms in Innertown, as "the entire land … is irredeemably soured … by years of run-off and soakaway" (Burnside 2009, 10): woods are poisoned, animals are genetically mutated, human bodies develop "unexplained clusters of rare cancers" (10), and human brains are afflicted with depression and madness. The plant is described as simultaneously "beautiful," "frightening," and (in a deliberate anthropomorphism) "sad" (62), and its mysterious agency is central to both the overall gothic affect and the thriller plot of the novel, as the plant is the site of the mysterious disappearance and presumed murder of several teenage boys. Sites of disappearance coincide with sites of toxicity, which prefigures similar constellations in recent Netflix-series like *Stranger Things* and the German *Dark*—works that, like *Glister*, articulate general questions of human environmental (ir)responsibility with specific histories of human violence.

Glister has been interpreted as exemplifying a new materialist aesthetic. For Seril Oppermann (2016), it is a novel "about how densely bodies and ecosystems are interrelated in ominous toxic kinship" (31). Foregrounding the agency of nonhuman elements like molecules and waste, *Glister* showcases the "naturalcultural dynamics of human-nonhuman existence" (30) and exposes "the hazy nature of boundaries between the social and the scientific, technology and morality" (31). So far so entangled, and it is true that the novel underlines the enmeshing of human and nonhuman matter, as when Leonard, the main narrator of the story, says that "[w]ith every breath [he] take[s] the world into [his] lungs … all the traces and smears and soot falls, all the threads of copper and nickel and 2,4,5-T … the future is written in our blood" (Burnside 2009, 70). Confronted with "sudden nosebleeds, numbness in [his] fingers, swollen knuckles, bleeding gums, gut pains," Leonard realizes that "there's something in [him]. Lurking. Some chemical trace, some cancer" (98). As these quotations show, *Glister*'s figurations of entangled matter are somewhat more sinister than the fairly upbeat "hybrid geographies" and "intermingling spillway" that the new materialism likes to evoke (Oppermann 2016, 32).

Nor does the new materialist notion of an interrelated web of distributed agents really exhaust the novel's evocation of the material world. The novel at times finds itself in tune with object-oriented ontologies as it underlines that some human and nonhuman objects remain resolutely inaccessible: one character, rushing to condemn the squalor of other people's lives, suddenly realizes that "[h]e didn't know what their lives were like" (23); confronted with his girlfriend-of-sorts' request to put a dying animal out of its misery "'[c]ause it's *suffering*," Leonard pleads non-access: "'How do you know?' … I have no idea what it's thinking, but I have no intention of killing it just because [my friend] is feeling squeamish" (140). Indeed, the emotive isolation of the different narrators and the radical disconnect between the characters is a crucial part of the novel's uncomfortable, gloomy mood.

The clearest resonance with object-oriented ontology, however, is in the novel's ultimate hyperobject: the Glister. A fairly obscure term—the word is more or less synonymous with "glitter" and "sparkle"—"the Glister" is introduced on the very first page; and while it is immediately associated with the plant site (which is vast in itself, and seems contiguous with the poisoned woods), it remains unclear whether these spaces coincide. It continues to defy localization and definition for most of the novel: a policeman finds one of the boy's bodies when entering "what looked like a little den among the trees," which, it turns out, "was really the first in a series of such closed spaces," ending in "a strange little bower" (26); it then turns out to be "some kind of machine, maybe a kiln, or a gas chamber" (216). It still defies description at the end of the novel: "the Glister. Which is what, exactly? A door? A portal?" (251). The Glister, it appears, serves as a site of connection, of relatedness, but it is itself a murky hyperobject of sorts that resists being mapped.

So what do we make of the novel's double alliance to materialist and object-oriented ontologies—ontologies that, as their proponents do not cease to point out, are rigorously incompatible? I think the most productive interpretation of *Glister* is as a work that *tests* these ontologies while *subscribing* to neither. Then we can see that it, for instance, critiques the ubiquitous relationality that it also evokes. In the early chapter entitled "Connections," the novel presents the arch-villain Brian Smith, whose distinctive skill is understanding that "*everything* was connected" (35), even people "when you see them as objects in the *fullest* sense of the world" (36). If this skill first asserts itself in a childhood obsession with puzzles, it is later sustained by the insight that what connects human and nonhuman objects is "the logic of money" (38). *Glister* thus establishes a clear link between materialist flat ontologies, in which nothing is hierarchically superior to anything else, and the role of money as a general commodity that makes everything exchangeable, including people's labor power. Add to that the fact that the novel's clearest visions of interconnectedness—"I can see everything around me in perfect, almost dizzying detail, but I can also feel how one thing is connected to the next" (129)—are drug-induced, and that the only form of collective action in the novel is an act of mob violence that kills an innocent man, and it becomes hard to read it as promoting a new materialist outlook.

That the characters refuse to blame anyone for the killing and describe it as "just one of those things" (186) seems to affirms Jane Bennett's (2010) belief that "[i]n a world of distributed agency, a hesitant attitude toward assigning singular blame becomes a presumptive virtue" (38). Yet what a new materialist or object-oriented perspective does not make visible is that the novel also sees this refusal to judge as a cop-out. The novel makes it very clear that the characters' ambivalent attachment to the plant and its poisoned environment is the unwanted result of economic decisions to abandon the plant and the community around it. Toxicity might be the effect of the thing-power of chemical substances, but it is also the result of the lack of caution exercised "by the Consortium, by the safety people, by all the powers that be" (10). This abandonment leaves the community no other option than to engage with the material reality of the plant, however toxic it is: "The Innertown wasn't a healthy place to live; the trouble was that, for most people, there was nowhere else to go" (30). What may seem like a wholesome sense of connectedness to the place is in fact a consequence of being left behind and of being stuck: "the people who live here are trapped, … they can't imagine any other life" (78). The numinous appeal of the plant is not a stable ontological feature of its plant-ness, it is an effect of socioeconomic violence: "if you want to stay alive … you have to love *something* and the one thing [to] love is the chemical plant … All we have is the plant" (60). For *Glister*, in short, not all differences are the same: class difference and the violence it entails are more crucial than either the relations between things or the abyss between objects. Coming to terms with the environmental crisis *Glister* evokes, then, is not only a matter of cultivating an appropriate mode of attunement, attentiveness, or responsiveness. As the next section elaborates, it requires operating on different scales.

Scale Shiftiness

In contemporary discourse, the notion of scale almost automatically evokes the infinitely large and the incredibly long. This is as true in environmental debates as it is in business lingo. When the business press talks about "scalability," it means the capacity of businesses to grow and to manage rising demand. Issues of scale pertain to the challenge of "scaling up" and of dealing with increased sizes. In the Anthropocene context, size matters too: the term names a moment when the sum of individual actions has come to affect the planet as a whole, and when the fallout of our actions resonates in faraway futures, rather than on the more manageable timelines on which we normally track human actions (days, weeks, lifetimes). Yet this "upscaling" does not mean that customary frameworks lose their hold on us: notions like "deep time" (a geological time that counts in billions of years) and "slow violence" (pointing to the gradual attrition of life worlds rather than to punctual disasters) are not absolute substances, but derive both their rhetorical force and their analytical significance from their contrast to the conventional extensions and rhythms of human life. Scale, in other words, is a *relational* notion: it names the *ratio* between different size domains. The Anthropocene is then not simply an outright "upscaling" of human life, but is better understood as a critical moment when size difference becomes a matter of concern. If we tend to take the rhythms, speeds, distances, and sizes of the human world for granted, the Anthropocene reminds us that the human scale is only one (set of) domain(s) among many.

Literature has always explored the tensions between the human mesoscale and different micro- and macroscales. One of the most famous examples is Jonathan Swift's 1726 *Gulliver's Travels*, which sends its protagonist to Lilliput, where he lives among tiny people, as well as to Brobdingnag, where he resides among giants over 60 feet tall. Such scalar mismatches would gradually disappear from the literary mainstream with the eighteenth-century rise and later consolidation of the novel, a form dedicated to the serious depiction of everyday life, and thus confined to the ebbing rhythms of human time (McGurl 2017). The imagining of macroscales was increasingly assigned to the genre of science fiction, as in Olaf Stapledon's 1930 work *Last and First Men*, which covers a sequence of 18 different human species in a span of 2 billion years. As scale has become a concern that also affects everyday life, it has increasingly come to invade the province of realist fiction. Think, for instance, of Karl-Ove Knausgaard's *My Struggle* project (2009–2011), which offers a 3,500-page exploration of even the minutest experiences, memories, reflections, and perceptions of the narrator. The scalar mismatch between the vast scope of the project and the infinitesimally small focus of the novels' mode of notation shows how Knausgaard's project unravels the customary rhythms and patterns of realism, and it helps explain both the fascination and the irritation the books have generated.

Don DeLillo's *Point Omega* (2010) offers a different approach to scalar variation. *Point Omega* is a decidedly slim work of some 120 pages, not only in comparison to Knausgaard's gigantic undertaking, but also to DeLillo's 1997 masterpiece *Underworld*, which connected different human-scaled narratives to provide an authoritative account of American postwar life. After its opening section, *Point Omega* abandons New York, the center of *Underworld*'s sprawling universe, and turns to the slow, eventless temporality of the desert, a space so indistinct it is either "the Sonoran desert or maybe it was the Mojave desert or another desert altogether" (DeLillo 2011, 25). Human geography does not matter, as this is the province of geological time—of "the protoworld … the seas and reefs of ten million years ago" (25). This is "deep time, epochal time" (91), in which human concerns, such as the disappearance of one of the characters, which is hardly investigated, "are overwhelmed by landscape" (82). The novel connects human life to geological and cosmic processes, and in this way also opens up the infinitesimal gaps that underlie our everyday functioning. At its beginning and its end, the novel evokes Douglas Gordon's *24 Psycho*, an art installation that slows down Hitchcock's film *Psycho* so it takes up 24 hours, the time it takes the earth to complete one rotation. This shift from the time of human entertainment to that of cosmic repetition robs the film of all suspense, and instead trains the viewer "to feel time passing, to be alive to what is happening in the smallest registers" (7). Projecting only two frames a second, the installation foregrounds the interstices of time—what *Point Omega* calls "submicroscopic moments" (21)—that are neutralized in ordinary perception. These infinitesimal interstices have as much of a disruptive impact on the human scale as the invocations of cosmic vastness that we find in the novel's central narrative. By destabilizing the human mesoscale, literature's scalar experiments participate in the same revisionary project that motivates the different materialisms, posthumanisms, and ontologies I discuss elsewhere in this chapter.

Complex engagements with different scales *simultaneously* offer a promising way to find a form for the Anthropocene. If Dipesh Chakrabarty in 2009 still noted that "[t]o call human beings geological agents is to scale up our imagination of the human" (2009, 206), he later appreciated that such scaling up is not enough, and that the real challenge is "think[ing] of human agency over multiple and incommensurable scales *at once*" (2012, 1, italics mine). Zach Horton (2017) usefully notes that the Anthropocene is less about the discovery of new scales than about the human's "confrontation with itself as a trans-scalar entity" (35). The illusion that we are monoscalar entities has made it possible to obfuscate the reality of scalar difference and to conjoin different scales within a single medium in a process of "scalar collapse" that, for Horton, characterizes both colonization and scientific rationality (36). Seeing human life as trans-scalar is then also a way of avoiding the imposition of one particular perspective on the complex environments that human action must negotiate. It makes it possible to appreciate that the new materialist injunction to be attentive to matter, for instance, cannot simply be scaled up to a planetary program to address the energy crisis, as such a solution demands tactics, campaigns, and decisions that may need to strategically bracket such attentiveness. Different size domains demand different approaches. In a curious 1926 essay entitled "On Being the Right Size," biologist J.B.S. Haldane notes that even Swift's Brobdingnag's giants are physically impossible: being "not only ten times as high" but also "ten times as wide and ten times as thick" as ordinary humans, the cross section of their bones (which determines the body's carrying power) could never support them, so they "would have broken their thighs every time they took a step" (1956, 952). It is no coincidence the human body is the size it is, and for Haldane, this also goes for human institutions: just as the Greek model of democracy can't simply be scaled up to a country the size of the UK, so socialism, Haldane surmises, might be extended to "Belgium Ltd or Denmark Inc.," but probably not to larger scales (956–57). Examples like these show that, when we talk about the human subject of the Anthropocene, scale shifts are marked by jumps, glitches, and discontinuities, not by smooth processes of zooming in and out.

As users of Google Earth and Google Maps, we all are familiar with the experience of frictionless scalability: these apps allow users to scan the globe and zoom in on any location of their choice. These apps' effortless performances of magnification and resolution afford their human users an experience of visual mastery and sovereign mobility. Charles and Ray Eames's 1977 short film *Powers of Ten* offers an early example of such zooming power. The film opens with a neatly composed picnic scene in a park, before the camera begins its vertical ascent at an ever increasing speed until it is 100 million light years away from the picnic scene. At this point, it begins its descent that returns to the park, only to collapse into a cell in one of the picnickers' hands and to further descend to the level of molecules, atoms, and beyond. The dispassionate seriousness of the film's voice-over reinforces the unperturbed performance of the camera zoom. It concludes the film by saying that "our journey has taken us through 40 powers of ten," instilling a sense of control over the different dimensions of reality (Dorrian 2011). Such a fantasy of scalar sovereignty is also at work in geoengineering and nanotechnological imaginaries. Yet there is also a more sinister dimension to the film, which hints at a more disturbing discontinuity underlying its visual virtuosity and impassive voice-over. These achievements cannot wholly neutralize the unease of seeing an emphatically human-scaled scene of leisure and happiness shrink in the face of a nonhuman power of vision that, on its way down, barely stops to register human bodies as it infiltrates them. The film displays an eerie power of abstraction and penetration. Even as it fosters illusions of technological mastery, it simultaneously qualifies the human scale as a powerless target of technological violence. In that sense, we can compare the film to the *Blue Marble* photograph I discussed in the previous chapter, in which human life is similarly rescaled by a technology that also intimates the power to annihilate it.

Fantasies of infinite scalability don't account for the scalar complexity of the Anthropocene world. Scale, Derek Woods (2018) notes, "is not a linear or zoom-like shift from big to small" (502), but is inevitably beset by what Woods calls *scale variance* (Timothy Clark uses the term *scale effects*). It is misleading, for instance, to see Timothy Morton's hyperobjects, which are massively distributed in time and space, as models to think of the object world at other scales. While it makes sense to say that things like Styrofoam or climate change (to name two exemplary hyperobjects) are so vast and unlocalized that they withdraw from human access, that is not necessarily the case for our relation to everyday objects. Similarly, the monism of the new materialism and the flat ontology it shares with Triple O fail to imagine a truly pluralist ontology that distinguishes different scale domains (Woods 2017, 217). These theories' blindness to scale makes sense in light of the customary association of scale jumping with power and control (Clarke and Wittenberg 2017, 11–12), but it leads to an impoverished account of how the climate-changed world works. There are "ontological rifts among size worlds" (Woods 2017, 210), and a failure to address those rifts overlays a world marked by discontinuities and thresholds with a smooth vision of openness, connectedness, and continuity.

When we abandon the notion that the subject of the Anthropocene is a scaled-up human, we can see that human action, at the planetary level, is thoroughly entangled with nonhuman forms of agency. The subject of the Anthropocene is not an inflated human, but, in Derek Woods's (2014) terms, "the sum of terraforming assemblages composed of humans, nonhuman species, and technics" (134). Factoring in scale variance is necessary to capture the distributed agencies that co-constitute the Anthropocene world. For Timothy Clark (2015), the Anthropocene is essentially "an emergent 'scale effect,'" as it interacts with nonhuman processes when it crosses a particular point of no return: "at a certain, indeterminate threshold, numerous human actions, insignificant in themselves (heating a house, clearing trees, flying between the continents, forest management) come together to form a new, imponderable physical event, altering the basic ecological cycles of the planet" (72). The different theories I survey in this chapter attune us to these different agencies, but it requires what Woods calls "scale critique" to emphasize disjunctions, incommensurable differences, jumps, and discontinuities and to develop a properly complex picture of "what forms agency takes and which mediators entangle it" (2014, 140). Such a picture will invariably combine cozy entanglements with more troubling disruptions and rifts.

Storied Matter in the Anthropocene Scriptorium

This chapter has shown that object-oriented ontology and new materialism return critical attention to the nonhuman world and underline that the Anthropocene is composed by the multifarious interactions between human and nonhuman agents. Still, our discussions of scale and of John Burnside's *Glister* have suggested that critical theory's nonhuman turn needs to be complemented with a consideration of discontinuities and tensions that cannot be neutralized in flat ontologies or in fantasies of benign interconnectedness. A full account of the Anthropocene world needs to encompass vertical difference and disconnection. In this final section, I want to suggest that imagining the interactions between Anthropocene agencies in terms of reciprocal storytelling and dialogue, as ecocritics Serenella Iovino and Serpil Oppermann propose, misses out on the power differentials, the violence, and the long-lasting and irrevocable consequences that also charactize these interactions. Writing, I think, offers a more promising model.

In Iovino and Oppermann's (2014) *material ecocriticism*, the last critical development I discuss in this chapter, storytelling provides the key metaphor for describing the way objects interact. For them, all matter is "storied matter" (1). All things are "undeniably expressive" and "have their own stories to tell" (Oppermann 2018, 9), and this turns the world into "a site of narrativity, a storied matter, a corporeal palimpsest in which stories are inscribed" (Iovino 2012, 451). For Iovino and Oppermann, material processes of interaction are deeply entwined with semiotic processes of meaning-making, to the point where the distinction between the two collapses—what they call "the porosity of biosphere and semiosphere" (Iovino and Oppermann 2014, 5). They write that "human and nonhuman players are interlocked in networks that produce undeniable signifying forces" (2). And when material interactions can be imagined as producing meaning, it is only a small step to imagine ecological interactions as cosmic conversations: Iovino and Oppermann write that "the world's phenomena are segments of a conversation between human and manifold nonhuman beings" (4).

This image of the world as a polite dinner conversation may not be fully adequate for rendering the upheavals of the Anthropocene world. Interestingly, Jane Bennett describes her own vibrant materialism as an attempt "to raise the volume on the vitality of matter" (2010, 10), an image that presents our engagement with the nonhuman world as a matter of amplification and noise management, rather than as a polite dialogue free from distortion or interference. In contrast, Iovino and Oppermann's ecological storytelling imaginary is conspicuously liberal, as it projects a fantasy of an infinitely patient dialogue between tolerant and generous conversation partners onto the nonhuman world. We can find comparable liberal imaginaries at work in both object-oriented ontology and the new materialism. In object-oriented ontology, as we saw, objects have the (all too human) capacity to withdraw from their relations to other objects and to introvert themselves. What is this if not a liberal vision of privacy, a belief that the core of our personal lives will be immune from interference? In Jane Bennett's vibrant materialism, all agents, whether they are humans, electric transmitters, or the fatty acid Omega-3, have the capacity to freely choose to engage, or not engage, with other things, as if they were happy liberals exercising their freedom of choice. As Thomas Lemke has noted, Bennett's extension of the category of the actor to the nonhuman world "still buys into the liberal concept of agency that sees it as a property of individual entities, focusing on will, freedom and choice" (2018, 41).

Such projections of liberal fantasies of privacy, of freedom of choice, and of polite conversation may not capture the realities of the Anthropocene world. This is a world where actions leave irrevocable traces, where some consequences cannot be undone, and where understanding requires less attentive listening than canny deciphering. These dimensions are perhaps less adequately represented by scenarios of interactive storytelling than by a model of writing and inscription. The activity of writing can easily be extended to nonhuman agents: in the Anthropocene, they leave traces that our technological instruments are increasingly capable of reading. And the Anthropocene is not only a topic of human writing, it is also defined by the fact that humans leave their traces in the geological record. These traces cannot be apologized for or quietly forgotten the way we can get over a bad story: they can only be read and reckoned with. Unlike storytelling, which can count on the continuous feedback that face-to-face interaction provides, writing takes place in the absence of its addressee and in ignorance of its own impact. This is also the case for the actions that cumulatively have led to climate upheaval. Writing is probably a more revealing model for Anthropocene agency than storytelling is: it captures dimensions of human responsibility, violence (as there is no inscription that is *not* the displacement of other matter), permanence (however unwanted), and discontinuity. Indeed, the juncture of literature and the Anthropocene is perhaps less a matter of uninterrupted storytelling than of a world imagined as an encompassing *scriptorium*: a world made up of the traces left by humans and nonhumans alike, and a world that our increased powers of reading do not allow us to leave unread.

Let's return to *Glister* one last time. In this novel, the way toxic waste affects the bodies and brains of people is less easily figured as a storytelling exchange than as a process of violent material inscription; the chemical waste participates less in the composition of what Oppermann (2014) calls an "epic of being" (30) than in the proliferation of death; it is less a matter of free expression than a lethal mode of writing. A story about the disappearance of young children, *Glister* is also a novel about the illusion that things can vanish without a trace: "No mark, no clue, no sign of a struggle, no note, no stain on the air at the point where he turned and walked away—if, as the adults tell us, he *chose* to go" (2009, 67). The novel here cannily links the illusion of tracelessness to the idea of freedom of choice, only to declare both notions obsolete. The novel criticizes the pretense of innocence and deniability by linking humans' constitutive role in planetary cataclysm to, precisely, writing: it is less a whodunit than an exercise in acknowledging a responsibility that we already know about, as "it's part of the plan of the story that you know already because these things are already written" (247). The existential situation the novel elaborates is that of the human in the Anthropocene. We are always already guilty, even if we never intended any wrong: "It's unforgivable not to know where they are, even if it's impossible to know" (250). Human responsibility is "already written," and attempts to address that responsibility only add more writing; they can never erase earlier inscriptions. The Anthropocene scriptorium does not offer the release or relief of a good story: it is marked by the impossibility of erasure, of forgetting, and of forgiveness, and by the compulsion to keep writing. In the Anthropocene scriptorium, it is always already too late for denial. The next chapter focuses on the notion of Anthropocene memory to explore in more detail what this injunction to remember looks like.

Part 2

Anthropocene Temporalities

4 Dominations

Anthropocene Angels

The Anthropocene warps our apprehension of time. It opens vast temporal expanses as it situates even the minutest human choices, such as that between driving a car or walking, in geological deep time. It turns the future into a site of dread and diminishment rather than rapt anticipation. The past is no longer a stable historical ground and morphs into a restive archive of stored actions that may one day prove our undoing as a species. The Anthropocene present, then, is a palimpsest of (often only partly legible) crisscrossing forces that do not provide a clear point of orientation to navigate the complexities of planetary life. This temporal disorientation is the organizing concern of the second half of this book, and it complements the emphasis on the mostly spatial relations between texts, disciplines, subjects, and things in the first half. It would be a mistake to see this temporal turmoil as radically unprecedented: industrial progress was always shadowed by an awareness of environmental devastation (Bonneuil and Fressoz 2016), and modern life, as it was typically lived in the buzzing metropolis, has often elicited a giddy sensation of clashing temporalities. Indeed, the idea that time is streamlined in an orderly progression from past over present to future has always been an ideological imposition. The cultural theorist Walter Benjamin (2006), in his classic "On the Concept of History," analyzes this quintessential modern myth of progress as "a homogeneous, empty time" (395) in which new experiences are neatly added in a linear and continuous fashion to the already accumulated data of the past.

Benjamin saw it as the task of the historian to insist on discontinuities, repetitions, and shocks that upset such a neat accretion and that instead point to "secret agreement[s]" between past and present (390). For Benjamin, this was essentially a task of developing an ethics of memory: such an ethics designs a relation to history that is "capable of fanning the spark of hope in the past" by remembering the violence and suffering on which reigning hegemonies are built, but which they work hard to forget (396). Benjamin's emblem for such an ethics of remembrance is the figure of the angel of history: keeping its face turned to the past even while the modern myth of progress propels it into the future, the angel contemplates the dismal record of the past as "one single catastrophe" and repels strategies to erase that record (392). It is no overstatement to say that Benjamin's angel has, in the last few decades, become an icon for a prominent tendency in critical thought that has mined the past for injustices and disasters in the hope of achieving a more equitable future. In the interdisciplinary field of memory studies, which has in recent years increasingly liaised with the environmental humanities, such a commitment to uncovering and undoing past wrongs has become programmatic. Memory studies wants to repair the erasures of "homogeneous, empty time" and to restore the temporal complexity of the present. In that sense, it has an elective affinity with the Anthropocene. This chapter traces that rapport in more detail by showing how the current environmental crisis affects memory and how the study of memory, in its turn, can enrich our understanding of the textured temporality of the present.

A complex and layered apprehension of time is not unique to the Anthropocene, but the turbulence of the Anthropocene present is yet different from previous iterations of temporal turmoil. As Jennifer Wenzel has argued, the temporal shifts associated with imminent climate dereliction and planetary exhaustion "destabilize the straightforward, secular assumption that pasts and presents have futures, that things just keep on going, that time and history keep unfolding" (Craps et al. 2018, 502). We now know that they will not. If Benjamin already knew that modernity's narrative of progress all too conveniently forgot the violence on which it was built, this narrative is now, Wenzel writes, "confounded once and for all by a future utterly different from that which fossil fuels once promised" (Craps et al. 503). The anticipation of a disastrous future also cancels the value of our memories of the past, as our past experiences were premised on a future that will not materialize: "Our memories of the future will have turned out to have been all wrong" (503). Interestingly, Benjamin's "On the Concept of History" already announces such a radical temporal upheaval. Near its end, the text quotes "a modern biologist" observing that "[i]n relation to the history of organic life on earth ... the paltry fifty millennia of *homo sapiens* constitute something like two seconds at the close of a twenty-four hour day" (2006, 396; Baucom 2012, 10–12). If the angel of history famously condenses a whole human history of violence into one pregnant moment when it sees "a chain of events" as "one single catastrophe which keeps piling wreckage upon wreckage" (392), this passage underlines that the history of human life is merely a moment in a much vaster history that is not merely human (263). It introduces organic and geological scales that complicate the mandate of memory studies and enjoins it to consider nonhuman forces in its memory work.

The second half of this book is organized around three rubrics that highlight the temporal disturbances that characterize the Anthropocene. This chapter treats the question of Anthropocene memory under the rubric of "dominations" to foreground the persistence of past injustices in the way the Anthropocene afflicts different communities. The next chapter is entitled "emergencies" and underlines the curious temporality of inhabiting a present when it is perhaps already too late to salvage the future. The last chapter, finally, uses the notion of "residues" to survey a number of engagements with the insight that human life will be reduced to a mere fossil in the planetary future. The series dominations/emergencies/residues remixes another triad of terms: Raymond Williams's famous distinction between "dominant," "residual," and "emergent" dimensions of the present. My variation on Williams's influential categories aims to emphasize that the present frenzy is continuous with, even if it intensifies, the disturbances Williams saw as endemic to social life.

Williams (1977)—like Benjamin, writing in a Marxist tradition—aimed to "recognize the complex interrelations between movements and tendencies both within and beyond a specific and effective dominance" (121). Williams defines the *residual* as those elements from the past that are still "an effective element of the present," and not simply archaic (122); "the actively residual," for Williams, is "alternative or oppositional," and not "wholly or largely incorporated" (122–23). The *emergent*, for its part, names those elements that point beyond the dominant: it indicates that "new meanings and values, new practices, new relationships and kinds of relationship are continually being created" (124). Just as the residual can be neutered when it is categorized as inoffensively archaic, the emergent risks being rendered harmless when it becomes fully incorporated as part of a self-updating dominant culture. It is crucial, then, to resist such neutralization by the dominant and maintain the residual and the emergent as irreducible dimensions of experience. In that way, Williams notes, local sites of the residual can even morph into emergent forces—a possibility that calls for a particular ethics of memory.

The residual and the emergent name two kinds of experiences that resist the false homogeneity of the dominant culture, and that turn the historical ground on which we stand into a treacherous and explosive terrain. Such experiences are still in search of a form, as they are "active and pressing but not yet fully articulated": emergent culture, Williams writes, "is never only a matter of immediate practice; indeed it depends crucially on finding new forms or adaptations of form" (126). It is a measure of the insistence of non-dominant elements that they fit uneasily in existing categories and forms, and that they often generate "an unease, a stress, a displacement, a latency" when we confront them with dominant (but, perhaps, soon to be defunct) forms (130). In the rest of this book, I take my cue from Williams's dynamic triad to describe the peculiar temporality of the Anthropocene in new forms and categories. The rubrics of dominations, emergencies, and residues remix Williams's terms for capturing the unsettled present and update them for a situation in which social life is entangled with biological and geological entities.

Anthropocene Memories

Approaching Anthropocene memory through the lens of *dominations* highlights that the present is marked by the persistence of histories of domination, of engrained inequalities, and of legacies of environmental injustice. If we extend the remit of memory to nonhuman actors and environmental disasters, an awareness of lingering domination stresses that we should not overlook the crucial role of human perpetrators, victims, and bystanders, and forget the disproportionate role of man-made structures such as capitalism, technology, and colonialism in natural violence. Benjamin's reference to the "paltry fifty millennia" of human life risks diminishing the role of human actors, while what is needed is, in Rick Crownshaw's words, "reading simultaneously within multiple and contradictory frameworks" (Craps et al. 2018, 502) and "address[ing] the humanist *and* posthumanist enclosures of memory" (Crownshaw 2017b, 246). In the context of memory, also, a careful calibration of the role of different actors is called for, and this means that human responsibility needs to be preserved even as agency is distributed across human and nonhuman participants alike. Since the beginning of the millennium, the study of memory has broadened its focus from national contexts to the transnational, multidirectional, cosmopolitan, and global drift of memory. But as I explained in the second chapter, the Anthropocene inaugurates a further shift beyond global to *earthly* and *planetary* dimensions. Let me briefly explain how this change affects four essential constituents of memory.

First, *it extends the objects of memory by also including nonhuman agents*. Anthropocene memory does not limit its concerns to human victims, but also factors in damage to lifeworlds, destroyed landscapes, and suffering animals. In the two literary works that I touch on in this chapter, nonhumans are objects of mourning: an Indian reserve in Thomas King's *The Back of the Turtle*, and Aboriginal habitats in Alexis Wright's *The Swan Book*. Nor is Anthropocene memory blind to the role of natural forces in bringing about, rather than undergoing, calamities. Violence can also be environmental, even if natural disasters are always partly man-made: think of how Hurricane Katrina, which devastated New Orleans in 2005, disproportionally affected African American communities because of decades of poor planning, inadequate provisions, and institutional failures. This differentiated geography of vulnerability was a consequence of human actions, and it amounted to a case of environmental racism. The broader scope of memory also entails an awareness of different scales, as, in Rick Crownshaw's (2017b) words, "cultural memory studies must track emergent causalities, ad hoc assemblages of agentive matter, and mutating patterns of change in predictable and unpredictable, calculable and incalculable ways" (245). Rosanne Kennedy has referred to the imperative to remember nonhuman actors without overlooking the human dimension as a form of "multidirectional eco-memory"—a mode of memory as attentive to difference as to entanglement (Craps et al. 2018, 506). Rob Nixon's (2011) notion of "socioenvironmental memory" captures a comparably broad range of scales and agents (24).

The object of Anthropocene memory, in short, is not simply a static object, but part of a complex process that also includes the subject doing the remembering. A second change is that *the range of subjects and media of memory are radically extended*, as the Anthropocene affects the conditions under which cultural knowledge is archived and remembered. If we typically think of the subject of memory as human, and of the media of memory as material carriers, the Anthropocene reimagines the whole planet as an archive of human actions (what I referred to at the end of the previous chapter as a vast planetary *scriptorium*): think of carbon dioxide levels in the atmosphere or the chemical make-up of sea and soil, which can be read as a durable record of human activity. As Claire Colebrook notes, "the inscriptive event of the Anthropocene is an *extension* of the archive, where one adds to the readability of books and other texts, the stratifications of the Earth" (Craps et al. 2018, 507). Nor is it the case that the planet only carries the imprint of *human* actions: it is also a witness to a geological history that long precedes "the paltry fifty millennia of *homo sapiens*." Bronislaw Szerszynski (2019) has proposed "thinking of the earth as something that remembers and forgets" (221). The combination of lateral tectonics, a buoyant continental crust, and the water cycle makes the earth a medium conducive to inscription and erasure, and contemporary technology has increasingly made it possible to open that terrestrial archive. Of course, technology is also making possible the wholesale destruction of the archive, and in that respect, the Anthropocene can also be seen as triggering a massive process of forgetting (233). Anthropocene memory resists this threat to the planetary archive. Ursula Heise (2016) has studied the encyclopedic thrust of global digital biodiversity databases that aim to preserve traces of species threatened with extinction and that function as a kind of "ecological epic" (15); Ben De Bruyn (2020) has shown that contemporary fiction functions as a repository for vanishing animal sounds in an age of sonic depletion. In these examples, digital archives and literary forms become visible as media of more-than-human memory.

The Anthropocene not only affects the objects, subjects, and media of memory, it also, and more counterintuitively, *alters the very temporality of remembrance* (and this is a third crucial change). Because the consequences of planetary destabilization are unevenly distributed, some constituencies are already living—or indeed, *already mourning*—a reality that more privileged communities fear will become part of their future. Jennifer Wenzel has pointed to "a strange inversion of colonial-era developmentalist progress narratives," as today the Global South is no longer supposed to catch up with the metropolitan North, but rather prefigures the horrors and devastations that will come to afflict even sites that have so far successfully outsourced the consequences of their actions (Craps et al. 2018, 504). The result, Wenzel writes, is "an inverse colonial fear: that an ominous Third World present offers an image of the First World's future" (504). One consequence of such temporal jolts is that imagining the future and remembering the past are no longer strictly separate activities: the future is no longer a site of infinite possibility but a replay of disasters already happening elsewhere, so remembering past cataclysms is not very different from anticipating the coming storm. This means that future-oriented genres, such as science fiction and other forms of speculative fiction, are also doing memory work: imagining a climate changed future gives us the present as the past of an imagined future, as a reality that will one day be nothing more than a memory, even as the speculative future resembles the kind of impoverished and brutalized world we used to think of as our own past. The temporal mode of such post-catastrophe imaginaries is that of the *future anterior*—"the dramatization of that which will have been" (Crownshaw 2017a, 128). Speculative fiction, as Rick Crownshaw notes, "gives narrative presence to that which is subject to cognitive dissonance if not disavowal in the present" (129). The resources of memory, and of the anticipated replay of the past that makes up the Anthropocene future, can complicate customary ways of inhabiting the present and remind us that the present is not a stable ground so much as a treacherous and uneven terrain.

If the future is already someone else's past, this means that non-traditional archives can also serve as resources for dealing with the future. *The renewed relevance of non-traditional archives* is a fourth feature of Anthropocene memory. Not only have Indigenous communities already shown resilience in the face of environmental devastation, they have also devised future-oriented imaginaries that, even if they belong to the past, can be updated for the future. Shelley Streeby (2018) has shown how Indigenous people and people of color have historically led the way in climate activism, and she has demonstrated that speculative fiction has been a key resource in these archives of hope (112). N.K. Jemesin, whose *Broken Earth* trilogy is situated in a world marked by periodic geological upheavals (as I already discussed at the end of the second chapter), underscores that her post-apocalyptic world-building is rooted in the experiences of non-dominant constituencies: Jemesin notes how she draws

> a lot of material from a number of different experiences of oppression, like being closeted from queerness, or drawing from the Holocaust ... when you look at human history, it's full of Fifth Seasons [the series' name for intermittent cosmic disasters], full of apocalypses ... I wanted to draw a world that felt realistic.
>
> (Hurley 2018, 472)

Remembering past and ongoing destitution, then, is vital for developing a more adequate sense of the present. And here we are back at Benjamin's conviction that memory can retrieve "a spark of hope" from the past: in order to disturb the "prevalence of ideas of security, prosperity, liberty, and the instrumentalization of nature and freedom" in official Anthropocene discourses (Crownshaw 2017a, 128), a critical

Anthropocene memory can attune itself to the past as not only a source of mourning, but also a resource of hope. By opening archives of subaltern or Indigenous knowledge, memory may yet find "a secret agreement between past generations and the present one" (Benjamin 2006, 390).

The rest of this chapter homes in on different aspects of the altered memorial landscape. The next section ("Anthropocene Turtles") reads Thomas King's *The Back of the Turtle* to show how contemporary literature can engage Indigenous archives to broaden the scope of memory by including nonhuman agents and by encompassing different modes of memorization. The section on "Anthropocene Swans" pays attention to the warped temporality of Anthropocene memory: bringing together several perspectives on processes of anticipatory (or proleptic) mourning, I turn to the Indigenous Australian writer Alexis Wright's *The Swan Book* to show how the speculative imagining of Aboriginal futures serves as a strategy to preserve Indigenous traditions as well as an intervention in the settler-colonial present. "Anthropocene Wilderness" closes off this chapter on Anthropocene memory by teasing out resonances between the curious temporality I explore and altered conceptualization of nature conservations. In the field of conservation, "New Conservationists" have taken on the lesson that human and nonhuman histories are so thoroughly interpenetrated that the wilderness can no longer be preserved, or even remembered, as a radically nonhuman site. The two novels I discuss demonstrate how this modified understanding of the environment affects how nature writing remembers the wild.

Anthropocene Turtles: Mining the Indigenous Archive

Written by Thomas King, an author of Cherokee and Greek descent, *The Back of the Turtle* is an award-winning Canadian novel from 2014 that explores the aftermath of an environmental disaster known as the Ruin (it is also referred to as "That One Bad Day" [King 2014, 181]). An attempt to clear foliage by administering a herbicide named "GreenSweep" goes wrong and leads to 137 deaths in a small Indigenous community in British Columbia, which has turned into a ghost town by the time the novels takes off. The novel focuses both on the biotech company responsible for the disaster and on the event's impact on a number of Indigenous characters returning to the abandoned site and restoring a tentative sense of community. In this way, it deviates from the customary cli-fi template, which tends to adopt the perspective of a white male scientist-hero to accentuate its commitment to technological solutions (Irr 2017; Kaplan 2015). In this novel, the scientist responsible for the cataclysm spends the novel trying to reconnect to his Indigenous past, an attempt that the novel supports by exploring the viability of Indigenous storytelling. The novel not only highlights how Indigenous populations and nonhuman landscapes are affected by manmade environmental disasters; it also explores Indigenous resources for imagining a resilient response to destruction.

The Back of the Turtle has no illusions that disasters won't happen again: in the novel's present, Domidion (the biotech company) is involved in an incident with an earthen dam that has dumped toxic sludge into a river and threatens other Indigenous communities. The novel positions the remnants of Indigenous life in a quietly resilient community as a counter to biotech's work of obliteration: there is the story of Mara, a native artist who returns to the old house of her deceased best friend after trying and failing to settle in Toronto; there is the old and wise Crisp, who serves as a living repository of the local folklore; there is a boy named Sonny, who runs a dilapidated hotel and converses with his absent father, who seems to be an almost godlike figure; and there is, especially, Gabriel, the renegade scientist working for Domidion who holds himself responsible for the Ruin that, he finds out, also killed his estranged mother, sister, and nephew. At the beginning of the story, Gabriel comes to the Smoke River Reserve to commit suicide, berating himself for being "the author of all that destruction," a destroyer of worlds (King 2014, 337, 168). At the end of the novel, he is part of the improvised community of survivors, even if some of these survivors are, it seems, merely imaginary (which is entirely unproblematic for the novel's casual magic realism). The novel mostly resists what Caren Irr (2017) diagnoses as an all too frequent "romanticization of an underdeveloped periphery" in cli-fi novels that imagine Indigenous communities as "saturated with a spiritual depth that compensates for material deprivation and powerlessness": both Gabriel and Mara feel deeply alienated when returning to the village, Sonny has lost his father, and all characters are marked by trauma and loss. Rather than "a romantic conception of pristine, alien wilderness," *The Back of the Turtle* envisions "the natural world as integrated with sometimes damaging human activity" (Irr 2017), even if, as we will see, the ending of the novel comes close to celebrating Indigenous authenticity. As the last section of this chapter elaborates, remembering nature, in the Anthropocene, means remembering the entanglement of human and nonhuman forces.

The Back of the Turtle's relevance for Anthropocene memory is not restricted to its depiction of the loss and devastation environmental destruction spreads among Indigenous populations. It stages the complicated and contentious dynamics of Anthropocene memory in different ways. The novel shows how corporate environmental devastation operates through a campaign of obfuscation and obliteration of the past. The firm's PR team aggressively promotes self-made footage to replace documentary evidence of its past crimes. Its CEO, whose existential unraveling the novel tracks relentlessly, ends up in an existential crisis obsessing over the question whether he "[w]ill ... be remembered?" (480), as if acknowledging that the regime he exemplifies thrives on enforced forgetting and carefully maintained deniability.

Against this campaign of forgetting, the novel mobilizes several modes of memory. Gabriel has disappeared from his lab and his apartment is found full of inscriptions that refer to sites of manmade environmental disasters: nuclear and biological waste dumps, anthrax facilities, and Indian reserves used as a bombing range (23). These notations exemplify a compulsive mode of inscription that points to the traumatic persistence of a violent past. It is this trauma that informs Gabriel's desire, at the novel's opening, to end his life. His debilitating guilt, the novel shows, is the destructive flipside of the corporate campaign of forgetting. As we will see, the novel will marshal alternative modes of memory that help the characters in the novel overcome self-hate and despair. The landscape also serves as an archive, as it testifies to a history of violence that it refuses to metabolize. The devastated reserve, we read, is full of bones: "Almost everywhere he looked, everywhere he walked, there were bones ... [t]hey lay out on the ground where the creatures had died, one minute alive, the next minute dead" (403). The reserve is, quite literally, "a graveyard" (74, 454), or "an authentic Aboriginal Ghost Town" (99). The novel shows how the environment refuses to participate in the aggressive corporate campaign of forgetting: while Domidion thinks it has adequately disposed of a barge sent out "to dump a mountain of toxic waste and incinerated biohazards into the ocean" (18) when the boat gets lost at sea, the lost barge returns together with Chinese victims of the accident that sank it, as witnesses to a violence that resists being erased. The landscape makes it impossible to forget that environmental history is, in Benjamin's words, "one single catastrophe which keeps piling wreckage upon wreckage" (392).

Gabriel's traumatic guilt and the landscape's scars testify to the ineradicable traces left by environmental violence. However, *The Back of the Turtle* also stages a more resilient mode of memory that, in Benjamin's terms, qualifies it as an effort to fan "the spark of hope" that memory work can retrieve from the past. Throughout, the novel foregrounds the power of storytelling as a regenerative device. At the heart of the novel is a collective act of storytelling that rehearses an Indigenous creation story ("The Woman Who Fell from the Sky") that affirms the connection between humans and animals, especially the titular turtles. Importantly, the storytelling not only implicates the Indigenous characters: it forges a community that also includes the Chinese boat people. Nor is the Indigenous tradition the only archive the novel draws on, as it also, for instance, uses many biblical elements in its own fairly exuberant, often funny, and decidedly magic realist storytelling, and as it incorporates many colloquial elements and instances of onomatopoeia that reveal its ambition to approximate oral modes of interaction. Ultimately, then, the novel positions itself as a medium of memory that can contribute to what we have called (following Rosanne Kennedy) a multidirectional eco-memory: a memory that includes animals and the land as subjects, objects, as well as media of memory.

Pheng Cheah (2016) has eloquently argued for the capacity of literary storytelling to interrupt the "calculative management and appropriation of time" that characterizes capitalist globalization (and that Benjamin would call "homogeneous, empty time") (17). Storytelling, Cheah believes, opens a new world, a temporal horizon that cannot be subdued by capitalist forces. And just as Raymond Williams underscores that emergent and residual forces cannot simply be enlisted as entertaining elements of dominant culture, this "normative horizon that transcends present reality" cannot be neutralized as an innocuous "alternative" temporality (13, 31): it is a radically open-ended affirmation of life. As we have seen, in the Anthropocene, such a shift from the abstractions of the *globe* to the lived reality of a *world* must find a way to recognize the *earthly* entanglements of Anthropocene life. When *The Back of the Turtle* opposes the destructive modes of remembrance embodied by capitalist obfuscation and traumatic inscription with a form of ecological storytelling, this constitutes an effort to restore the world by remembering endangered modes of life and violent fits of destruction.

Anthropocene Swans: Mourning the Future

The Back of the Turtle's affirmation of creativity and connectedness offers a particularly hopeful inflection of Anthropocene memory. Yet hope is not the only option, as time for decisive climate action runs out and legacies of domination linger on. When the daily spectacle of environmental destruction tells us that the future we anticipated may never arrive, this can give rise to psychological distress. To describe this psychic pain, the Australian philosopher Glenn Albrecht (2005) has coined the term *solastalgia*, a neologism that evokes desolation, isolation, and an impotent longing for a lost state of solace. Solastalgia describes the psychological pain experienced when one's sense of belonging is undone by environmental change: it is "the 'lived experience' of the loss of the present as manifest in a feeling of dislocation; of being undermined by forces that destroy the potential for solace to be derived from the present" (45). If nostalgia (or homesickness) occurs when one *leaves* home, solastalgia is a situation where *staying* home makes you sick because the place that once was home is being destroyed before your very eyes, leaving you powerless and depleted. *Ecological grief* is a related term that captures feelings of anger, powerlessness, and depression in the face of an ongoing or anticipated destruction of the lifeworlds that sustain us (Cunsolo and Ellis 2018). It is no coincidence that both terms were developed in relation to the experiences of Indigenous populations, as these are often the first to experience the impact of climate change: their past and present is our future. A third term that reflects environmental distress is closer to the experiences of privileged constituencies who are not yet living the disastrous consequences of planetary deterioration on a daily basis. E. Ann Kaplan (2015) talks about "Pretraumatic Stress Syndrome," in which "people unconsciously suffer from an immobilizing anticipatory anxiety about the future" (xix). *Pretrauma* is exacerbated by the media's continuous bombardment of painful images, demoralizing stories, and depressing data, which induces a state of vulnerability and uncertainty. It is different from PTSD, in which the present is overshadowed by the reality of past events rather than by the prospect of future ones.

Solastalgia, ecological grief, and pretrauma: these are three dysphoric terms that capture the foreclosure of the future and the erosion of the present. While all three terms are beset by negative affect, they do not necessarily lead to immobilizing despair. In his book *The Uninhabitable Earth*, a terrifying vision of a warmed near-future earth, David Wallace-Wells (2019) argues that alarmism is the appropriate response to the deterioration of the planet, and the only one likely to break people's complacency. Kaplan too hopes that pretrauma might not leave audiences "passively terrified" but will instead help them "understand dystopian scenarios as warning humans of what they must, at all cost, avoid" (2015, xix). It has always been the mandate of dystopian fiction to warn contemporary audiences of dangers that, if we fail to address them, will result in the living hell (or the uninhabitable earth) that the dystopia depicts. Dystopia, like the post-catastrophe and post-apocalyptic fiction I discuss in the last chapter, typically serves as a "prophetic vehicle" that aims to preempt the horrors it depicts (Baccolini and Moylan 2003, 1). Dystopia's characteristic future anterior, Rick Crownshaw (2017b) notes, "dramatizes an etiology of the conditions that are imagined in the future but that are unfolding in the present of this literature's production and consumption" (244). When this etiology reveals destructive processes that are no longer reversible or shows that future calamities merely replay past and present forms of domination, as it often does in the context of the Anthropocene, dystopia's prophetic and preemptive potential is challenged. Engagement with the foreclosure of the future then easily shades into psychological distress and into debilitating modes of anticipatory or preliminary mourning.

Alexis Wright's *The Swan Book* (2013) showcases the temporal and affective complexities of dystopian fiction in the Anthropocene. Wright is an Indigenous writer and activist belonging to the Waanyi people, who live in the Gulf of Carpentaria in the North of Australia. The story is set three centuries after the 1788 colonial invasion of Australia, in the aftermath of environmental disasters, global resource wars, and disruptive migratory movements. The lake where the Indigenous group lives is reduced to a swamp, which serves as a dumping ground for "decaying plastic, unwanted clothes, rotting vegetable matter or slime that bobbed, *wanami* diesel slick" (Wright 2016, 36). It is also an "oasis of abandonment" for humans: the "Boat people" coming from overseas and the "truck people" hailing from elsewhere in Australia (56, 52; Sefton-Rowston 2017, 652). Environmental change has brought black swans, which used to reside in the South of the country, to the swamp. The setting of the novel, then, conveys a strong sense of *solastalgia*: its sense of place manifests as an affect of dispossession and diminishment. Indigenous life is typically characterized by a reciprocal relation of care between the land and its inhabitants, a relation captured in the notion of Country (Gleeson-White 2017 29). In the novel's climate changed future, that relation is beset by distress and harm.

The Swan Book not only accentuates the continuities between the diminished future and three centuries of colonial violence, it even more emphatically ties the future to recent and ongoing developments. The harassment of Indigenous populations clearly echoes the 2007 Northern Territory National Emergency Response, commonly referred to as the

Invasion, which upset Indigenous ways of life in response to unsubstantiated reports of child abuse (Johns-Putra 2018, 34). The main character, the mute girl Oblivia, becomes the victim of a gang-rape by a group of derelict youngsters, which indicates the incompetent handling of Indigenous populations the novel condemns. Oblivia's name, which evokes the temptation of forgetting and disconnecting from the past (Takolander 2016, 114), stresses that the aggression against Indigenous life is also an assault on memory. This threat first comes in the form of Bella Donna, a white European climate refugee who adopts the young girl, gives her her name, and fills her head with stories of white swans, as if to crowd out her native knowledge. Bella Donna is a clear allegory for the missionary zeal and cultural arrogance of the colonizer, who aimed to replace Indigenous cultural memory with an imposed British framework.

While contemporary cultural politics is officially committed to diversity and multiculturalism, the novel shows that it perpetuates this assault on Indigenous culture. The novel's main plot involves Oblivia's marriage to, later abandonment by, and subsequent flight from Warren Finch, Australia's first Aboriginal president. Warren is "post-racial. Possibly even post-Indigenous" (Wright 2016, 122): he does not believe in antagonism, but in assimilation into a global order premised on tolerance and human rights. Aboriginals are compelled to participate in this order, as a failure "*to change things by themselves for the future*" means that they have forfeited their "*right of sovereignty over their lives*" (232). Warren's voice becomes the only Aboriginal voice that is tolerated, and it displaces alternative and more confrontational instances of Indigenous advocacy (291). In the name of tolerance, the *dominant* culture perpetuates a history of domination that wages war on *residual* cultures so as to prevent their transformation into *emergent* collectivities. After forcing Oblivia to marry him, Warren orders the destruction of the swamp from which she hails and imprisons her in a tower in the city, where her life seems "to have coalesced into a stream of forgetting" (318). As Benjamin observed, the imposition of a homogeneous time of fungibility operates through the destruction of non-dominant memory.

The Swan Book does not share *The Back of the Turtle*'s cheerful confidence in human resilience and storytelling. Its insistence on relentless and intensifying environmental and societal deterioration is more likely to induce *pretrauma* than climate action. The novel depicts late-twenty-first-century Australia as "a dilapidated country in a dilapidated world" (266), and it refuses a vision of reconciliation; it breaks with a tendency in 1990s Indigenous Australian writing that aimed at a "reconciliatory literature" (Sefton-Rowston 2017). Chronicling the destruction of the Indigenous relation to Country, the novel is committed to salvaging a minimal remainder of that relation. This relation radically privileges nonhuman actors, such as the titular swans, but also other birds. The novel ends with a strange flash forward to a scorched planet marked by massive species extinction, when all the swans have deserted the swamp and only myna birds remain. These birds, it seems, have adopted Oblivia's role of speaking and writing the Country: they speak "the traditional language for the country that was no longer spoken by any living human being on the Earth" (329). The myna birds create "glimpses of a new internationally dimensional language about global warming and changing climates for this land" (329). The most horrifying aspect of this nonhuman speech is that it is not certain that there are human listeners left to hear it.

And still, *The Swan Book*'s posthumous imagination provides a measure of hope beyond hope. The story is framed in a way that leaves open the option that the whole narrative is generated in Oblivia's brain. Her brain is affected by a virus "manufactur[ing] really dangerous ideas as arsenal" (1). This destructive virus is the result of ideas that "had originated somewhere else on the planet and got bogged in [her] brain" (3) and undercut the possibility of developing a more viable relation to the land. The story we read is then the record of her "quest to regain sovereignty over [her] own brain" (4). The novel presents this quest as a form of geological inscription, as a way of writing the land: the girl is "writing stanzas in ancient symbols wherever she [can] touch," and she keeps on "writing the swirl language over the dust that fell on what the tree had witnessed in its lifetime" (7, 10). The novel's emphatically *lyrical* quality—its mixture of "music, poetry, lyrics, whispering spirits, myths, symbols, and interior monologue" (Sefton-Rowston 2017, 651)—then testifies to its ambition to write its way toward a new rapport between nonhuman environment and human action, and toward a reciprocal relation in which different actors write one another into existence.

I ended my previous chapter by proposing writing and inscription as more adequate literary models for the Anthropocene condition than storytelling. The *Swan Book* seems to agree. The text of the novel is not voiced, as Oblivia is mute. And in the multicultural politics that the novel depicts, her voice could only ever be neutralized and could never really make a difference. As Adelle Sefton-Rowston notes, the novel is "speaking in various echoes rather than one distinct voice" (651). *The Swan Book* declines the commitment to the powers of voice, world-making, and creativity that is on display in *The Back of the Turtle*. The emphasis is on the realities of domination and on intimations of a future that is no longer human. When the novel ends by looking back on the present from a disenchanted posthumous perspective, it constitutes the present as a future memory and it diagnoses solastalgia—or ecological grief, or pretrauma—as entirely appropriate postures.

Anthropocene Wilderness: After Conservation

Both *The Swan Book* and *The Back of the Turtle* mine Indigenous archives to explore how industrial and capitalist development has spoiled more symbiotic relations between the human and the nonhuman: through spectacular toxic spills and habitat destruction, but also through racialized forms of biopolitical governance and slow violence that are endemic to settler colonialism, and thus to the histories of Canada and Australia. The lifeworlds evoked in these novels are indelibly contaminated and compromised by human impact, but that does not mean that the novels automatically sustain the compensatory fantasy of a supposedly pristine environment that is free from human influence. Environmental thought has long been organized around the opposition between wilderness and civilization, and nature writing has played a crucial role in cementing that distinction. In the United States, for instance, the nature writing of Henry David Thoreau, John Muir, and Edward Abbey has helped naturalize a vision of the wilderness and civilization as two separate realms. This position has in its turn informed a conservationist ethos that aims to keep the wilderness intact and to restore decaying ecosystems. This mindset was phrased most memorably by American president Theodore Roosevelt, when he instructed, in relation to the Grand Canyon: "leave it as it is. You cannot improve on it. The ages have been at work on it, and man can only mar it" (qt. Marris 2015).

Such a commitment to the preservation, or even restoration, of the fading reality of an unspoiled wilderness is also a matter of memory. At the same time, the very idea of a domain divorced from human corruption is incompatible with the insistence on the entanglement of human and natural histories that is central to Anthropocene thought. Because the notion of the wilderness relies on an impoverished account of the relation between nature and culture, the "Old Conservationism" is increasingly being replaced by a so-called "New Conservationism" that sees nature as a territory to be managed rather than preserved intact. Nature is a site of responsibility, not of authenticity. Works like *The*

Back of the Turtle and *The Swan Book* maintain an ambivalent relation with this shift. By drawing on Indigenous archives, they make the point that what dominant settler experience constructs as an unpopulated wilderness has, in fact, always been the site of intense interactions between human and nonhuman actors. At the same time, *The Back of the Turtle* comes close to celebrating such primordial interactions as more authentic than the forms of life that civilization imposes. The novel ends when the reserve's ecosystem, which had earlier been destroyed by toxic waste, is about to be restored: fish, urchins, and anemones move in again, followed by birds and turtles. The role of humans in this scene is limited to "banging the drum, singing [a] turtle-bone song," "danc[ing] in the air and walking on water" (King 2014, 516–17). This celebration restores the unity between human and nonhuman life, but it arguably perpetuates, rather than challenges, the wilderness imagery that has often served as an alibi for colonial conquest.

In early environmental writing, wilderness was conceived as society's other: it was "nature in a state uncontaminated by civilization" (Garrard 2004, 59), an easily idealized realm divorced from technological and industrial pollution and corruption. Environmental historian William Cronon's classic essay "The Trouble with Wilderness" not only unsettles this distinction on conceptual grounds, it also points to the considerable social and political costs incurred by maintaining it. Cronon observes that the very opposition between wilderness and civilization is a construction *from within* civilization, and thus starts from a position of alienation from nature. The celebration of wilderness, Cronon (1996) notes, "has been an activity mainly for well-to-do city folks" (17). Glorification of the wilderness, Cronon continues, "embodies a dualistic vision in which the human is entirely outside the natural" (17). This allows societies to sidestep the difficult issue of how to design viable ways to live with and to develop sustainable relations to nature. Such a more responsible interaction with nature is of crucial importance: "To the extent that biological diversity … is likely to survive in the future only by the most vigilant and self-conscious management of the ecosystems that sustain it, the ideology of wilderness," Cronon notes, "is potentially in direct conflict with the very thing it encourages us to protect" (18).

This call for a more responsible management of nature informs contemporary conservation practices, which increasingly recognize the inevitability of human intervention because, as environmental writer Emma Marris (2015) writes, "[t]oday we can't withdraw without blood on our hands." Human action has fundamentally shaped the lifeworld of plants and animals, and preventing species from going extinct often requires further active monitoring and intervention. Conservation is a key context in which Anthropocene discussions over human responsibility, impact, and capacity need to be translated in concrete actions and practical decisions. Marris's (2011) insights in her book *Rambunctious Garden: Saving Nature in a Post-Wild World* resonate with those of so-called New Conservationists like Peter Kareiva and Joseph Mascaro and move beyond the "Old Conservationist" emphasis on preserving the wild intact. As Jamie Lorimer (2015) notes, such traditional conservation is "reactive" as "[i]t seeks to preserve a fixed Nature from modern, urban, and industrial Society by enclosing it in National Parks" (5). It also misrecognizes how thoroughly human cultivation has shaped the land for thousands of years, also in Pre-Columbian America.

For New Conservationists, the preservation and restoration of prehuman worlds have become obsolete projects in a post-wild world thoroughly shaped by human action. Saving the environment now requires removing or resettling species, deploying nonnative species, or supporting new ecosystems in humanized environments. A preservationist ethic insisting on purity and non-intervention merely puts a brake on adequate action. For New Conservationists, non-intervention is neither a goal nor a virtue, as the relevant choice is between good interventions and bad interventions, between effective and destructive wildlife management. Anthropocene nature cannot be left to its own devices at a remove from civilization; instead, as Jamie Lorimer (2015) notes, we need ways to find a place for "wildness at the heart of contemporary life" as part of a settlement that sustains rather than alienates life (11). Such a new conservationist trust in the capacity of humans to serve as responsible and competent gardeners maintains an ambiguous relation to the ecomodernist belief that we can "decouple" human well-being from environmental destruction. On the one hand, recent conservation practices, like ecomodernists, are committed to the human ability to manage environments that need intervention precisely because human mismanagement has made their survival precarious. On the other, when the "Ecomodernist Manifesto" notes that "decoupling" raises the possibility "that societies might achieve peak human impact without intruding much further on relatively untouched areas" (Asafu-Asjaye 2015, 19), the belief in eco-management shades back into the ideological construction of the wilderness that Cronon criticized and that contemporary conservationism has left behind. Instead of a sanitized decoupling between the wilderness and civilization, *The Swan*

Book ends with a rambunctious wasteland created by human mismanagement. Its final vision is of a posthumous world where the lake has become a "dry swamp," full of "waste" and "rusted junk" (Wright 2016, 297). The swans have been replaced by "ghost swans" and by myna birds singing "songs about salvaging and saving things" (297). The decision what to salvage from a history of domination and save for a climate changed future is an urgent one. The next chapter shows how literature can help bring that urgency into relief.

5 Emergencies

Emergency vs. Infrastructure

We can think of the warped temporal experience of the Anthropocene like this: even if a minimal scientific literacy suffices to *know* we need to act, it all too rarely *feels* like we are in a hurry. The temporal and spatial grammar of planetary crisis rarely takes a form that moves us from inertia to urgency: rather than an identifiable event in a particular location, planetary dereliction is a spatially and temporally distributed phenomenon. Rob Nixon's phrase "slow violence" has rightly become famous, as it captures the quietly devastating impact of processes that never really register as emergencies. Nixon (2011) defines slow violence as "a violence that occurs gradually and out of sight, a violence of delayed destruction that is dispersed across time and space, an attritional violence that is typically not viewed as violence at all" (2). Nixon's focus on (in)visibility is apt: whether we think of expanding pools of plastic, rising temperatures, or steadily diminishing biodiversity, these processes only gain visibility when they cross a particular tipping point and congeal into particular images: of desertification, of forest fires, or of the corpses of sea animals with stomachs full of plastics. But even if we *know* that the consequences of the deforestation in the Amazon or the heating of the Arctic will eventually come to upset our lifeworlds, and even if we *feel* the affective appeal of the powerful images, the long, murky, and intricate causal chains connecting dispersed moments and locations still complicate the translation of such awareness into a sense of urgency.

If the previous chapter zoomed in on the way the Anthropocene affects our remembrance of the past and the next chapter turns to Anthropocene anticipations of an uncertain future, this chapter homes in on the difficulty of inhabiting the present and apprehending the urgency of instantaneous action. Until about a decade ago, it was fashionable to hold that the modern interplay between past, present, and future had made way for an "eternal" postmodern present. In this extended present, we were condemned to enjoy a situation of perpetual sameness in which we found ourselves disconnected from the past and radical change seemed as undesirable as it was inconceivable. Such complacency is now a thing of the past: in the climate changed present, we know that the accretion of past actions is gradually emerging to become a threat to our established forms of life. *And yet, such emergence does not register as an emergency.* In the face of such invisible forces, affective blockages, and attention deficits, literature has its job cut out for it: it can mobilize its formal and thematic resources to help move the sluggish and obscure processes of the Anthropocene over thresholds where invisibility and indifference make way for urgent concern. Often, literature does this by imagining a near future when processes that have not fully emerged in the present will have surfaced and impacted everyday life. I turn to such dystopian and post-catastrophe futures in the final chapter.

In this chapter, I explore literature's capacity to amplify the barely audible background noise of contemporary life by focusing on the notion of *infrastructure*. Infrastructure—railroads, the electric grid, bridges, fiber optic cables—is a measure of the extent of human impact on the planet: for Jedediah Purdy (2019), we are "an infrastructure species," as "the material habitat that humans have created" currently amounts to "approximately four thousand tons of transformed world per human being, or twenty-seven tons of technosphere for each pound of a hundred-fifty-pound person" (88, 22). Yet infrastructure is also the boring, unglamorous, and ambient background of human life: it is supposed to do its invisible work while we are busy doing more exciting things, and we only really notice it when it breaks down. This is the conceit of a lot of near-future dystopias: they activate our desire for unglamorous things like transport and communication networks by presenting a world where these conveniences have disappeared. The tendency to take infrastructure for granted is intensified in the digital present, when "the material networks that make our seemingly immaterial systems work" are carefully erased in fantasies of weightless clouds and speedy wirelessness (Mattern 2013a). Infrastructure, in other words, brings into relief the tension between emergencies and everyday life; it highlights that only the sudden interruption of the everyday can generate a sense of urgency.

Yet infrastructure is not only something we take for granted: especially in the Global South, infrastructure works can be deeply aspirational—as when a road or bridge holds the promise of a more convenient or more affluent life (Boyer 2019). In the next section, I show how this tension between smug entitlement (on the part of Western elites) and eager ambition (for disenfranchised communities) makes infrastructure a key site for exploring the fault lines that, in Steve Mentz's (2019) words, "break up" and "pluralize" the Anthropocene. A passage from Jamaica Kincaid's *A Small Place* shows how literature engages differentiated experiences of infrastructure. The third section of this chapter shows how the brittleness of infrastructure has become increasingly inescapable after decades of neoliberal neglect. In the last few decades, relentless exploitation of the planet has gone hand in hand with a hollowing out of what philosopher Bonnie Honig calls "public things." For Honig (2017), shared spaces and provisions are constitutive of political life: public things like "sewage treatment plants and railroads" are vital because they "constitute us, complement us, limit us, thwart us, and interpellate us into democratic citizenship" (5). I turn to Karen Tei Yamashita's *Tropic of Orange* to explore how literature can mobilize the vulnerability and resilience of infrastructure—in this case, the L.A. freeway system—for the imagining of political hope. In the final section, I turn to the topic of energy, a key infrastructure that fossil fuel modernity has long taken for granted but that has now entered a sustainability crisis that calls for a swerve to less destructive ways of fueling life. My literary example comes from Ben Lerner's *10:04*, a work that leverages the vulnerability of the electric grid (revealed in the Manhattan blackout at the time of Superstorm Sandy in 2012) to intimate a changed future.

My literary examples focus on energy and transport infrastructures to unsettle the customary distribution between the spectacular and the invisible. Disturbing that distinction is a key literary strategy for highlighting the slow and subterranean processes that undermine contemporary life. At the same time, they articulate such an engagement with a half-hidden common world as a vital political concern. Belonging to different genres—postcolonial nonfiction (Kincaid), multicultural postmodernism (Yamashita), and post-postmodern autofiction (Lerner)—these works cumulatively testify to the literary project of inhabiting a present that, composed by the crisscrossing trajectories of different emergent forces, resists a sense of emergency. By redrawing the relation between everyday life and climate emergency, these literary works contribute to the imagining of what Stephanie LeMenager (2017a) has called "the everyday Anthropocene." This everyday Anthropocene eschews the gravitas of bold epochal statements, and instead lingers on "the present tense, lived time of the Anthropocene," on "what it means to live, day by day, through climate shift and the economic and sociological injuries that underwrite it" (225). LeMenager sees literatures of the everyday Anthropocene connect "individual, fragile bodies" to a collective "project of staying home and … *making* home of a broken world" (225–26). The works I discuss in this chapter uphold infrastructure as a crucial dimension of such everyday world-building.

Infrastructural (In)visibilities

The etymology of the term "infrastructure" gives us a clue to its appeal in discussions of the Anthropocene. Infra-structure means "below-structure," and the term refers to "the innards of a structure that are hidden by the structure's surface or facade" (Rubenstein et al. 2015, 575). Infrastructure, that is, is the invisible but decidedly material ("innards") dark side of manmade constructs ("structures"). It is what supports the phenomenal side of human enterprises: the steel and concrete underpinnings of cities, or (increasingly) the invisible communication networks that instantaneously transport data between users. Such "below-structures" are not supposed to force themselves into the foreground: they are "the invisible, forgettable ambiance in which the daily drama of modern life takes place" (585). As "the undergirding of modern societies, … they generate the ambient environment of everyday life" (Larkin 2013, 328). These formulations underline the *atmospheric* quality of the way infrastructure enters everyday experience: never as the focus of attention, only ever as a sustaining environment.

The current academic interest in infrastructure indicates that infrastructure has increasingly refused to remain merely ambient. For Dominic Boyer, "today's focus on infrastructure often appears as commentary upon infrastructural collapse, infrastructural decline and decay." After decades of neglect of public services and provisions, we are presently living the "incipient failure of neoliberalism to really deliver on its own promises" (Boyer 2014). A process of disinvestment and "slow erosion" has crossed a critical threshold as we see roads, energy plants, and libraries fall into disrepair (Rubenstein et al. 2015, 578). The "innards" of societal structures have stopped reliably radiating a pleasant ambiance, and insistently remind us of their metabolic function: the fact that human culture consists in the processing of nonhuman matter, and that infrastructure is a zone where human designs encounter nonhuman forces they can only partly control. Infrastructure is a site where human and nonhuman actors interact with one another, as Adam Rothsthein (2015) underlines when he defines infrastructure as "the underground, the conduited, the containerized, the concreted, the shielded, the buried, the built up, the broadcast, the palletized, the addressed, the routed". If this sequence resembles the object-oriented litanies I discussed in the third chapter, it differs from those evocative lists in insisting on (rather than bracketing) human mediation. The current interest in infrastructure, then, goes to the very heart of the Anthropocene: it points to the interface of the human and the nonhuman and reminds us that the interactions between them, like all encounters with objects, matters, and things, are an essentially political concern. Dominic Boyer (2019) has noted a "basic but also invisible codependence" between infrastructure and political power, as large-scale infrastructure projects help concentrate political authority in the hands of powerful elites (16). Infrastructures only *seem* apolitical, because, as Bruce Robbins (2007) notes, "they seem to constitute a minimum threshold, an earth-bound zone in which the large irresolutions of politics can for once be ignored and decisions safely left to the technocrats" (31).

The idea that infrastructure is invisible and only ever merely ambient is a fairly provincial one. It betrays a privilege that, as bridges come crumbling down and urban traffic becomes deadlocked, is increasingly untenable. In the introduction to an issue of *Modern Fiction Studies* on "infrastructuralism," the editors attribute the "inherent boringness" of infrastructure to two features. One is *scale*: infrastructure is so vast and sublime that it threatens to dwarf our everyday concerns, and this makes it expedient to just ignore it (Rubenstein et al. 2015, 575). The other is a matter of *desire*: because infrastructure is mostly public, it does not qualify as a commodity we want to own or consume. Infrastructure, in the words of Bruce Robbins, "is the object of no one's desire" (2007, 26). Uncontained and undesirable, infrastructure is just supposed to work without imposing any affective demands. While this position is prevalent in literary studies, anthropological studies of infrastructure have made more allowances for the aspirational dimension of works of infrastructure. In this context, the vast size of public projects is crucial to their desirability. Anthropologist Brian Larkin (2013) has remarked that "many studies that begin by stating how infrastructures are invisible until they break down are fundamentally inaccurate": if invisibility is one aspect of infrastructure, it is "only one and at the extreme edge of a range of visibilities that move from unseen to grand spectacles and everything in between" (336). Patricia Yaeger (2007) argues that infrastructure can become the object of dreams, and can even become "hypervisible" (16). Anthropologist Hannah Appel (2018) gives the example of Equatorial Guinea, a country with one of the highest investment percentages in the world, where the "infrastructure frenzy" saturates daily life and appeals to the population in visceral, sensory ways through "the endless thrum of jackhammers, bulldozers, and trucks too big for old colonial roads; the air full of cement dust that settles on skin and in mouths" (42–43). Indifference and undesirability are not the last words when we want to understand the complexities of infrastructure.

The literary critical insistence on infrastructural invisibility is in a sense unsurprising: the field of literary studies has traditionally been beholden to practices of unmasking and revelation that disassemble surfaces and facades to reveal the hidden depths of textual meaning. Yet as I showed in the first chapter, such practices of demystification have increasingly made way for readerly attitudes that aim to capture the specific textures of aesthetic and textual constructs and the whole range of our engagements with them. Brian Larkin (2018) remarks that "infrastructural inversion," "bringing what is background into the foreground," does not consist in an absolute shift from invisibility to visibility, simply because visibility and invisibility are not ontological features of objects but emerge "as part of technical, political, and representational processes" (186). The distinction between spectacular infrastructures and mundane ones, then, represents "different styles of visibility" (186) that can be described in all their variety rather than posited as a scaffold for heroic critical interventions.

To get a sense of how literature can contribute to such a critical description of the workings and the politics of infrastructure, I turn to a short passage in Antiguan-American writer Jamaica Kincaid's work of creative nonfiction *A Small Place* (1988). An angry indictment of the way Antigua, a small island in the West Indies, is being mistreated by incompetent leaders, the international tourist industry, and the legacies of the British Empire, the book's first section addresses an imagined tourist in the second person. Looking at the beautiful shore ("Oh, what beauty! Oh, what beauty!") through her hotel window, the imagined tourist pictures herself on the beach:

> You see yourself taking a walk on the beach … You see yourself eating some delicious, locally grown food … You must not wonder what exactly happened to the contents of your lavatory when you flushed it. You must not wonder where your bathwater went when you pulled out the stopper … Oh, it might end up in the water you are thinking of taking a swim in; the contents of your lavatory

> might, just might, graze gently against your ankle as you wade carefree in the water, for you see, in Antigua, there is no proper sewage-disposal system.
>
> (Kincaid 2000, 13–14)

The passage exposes the gap between rich western tourists, who travel to Antigua with the help of dependable transportation systems (the book opens with an airplane landing on the island's international airport), and the local population, who even lack the infrastructure to dispose of their own shit. For the former, mobility and clean water are taken for granted; for the latter, they are the promise of a more hygienic and more dignified life. Yet crucially, the passage also destabilizes that opposition: if the local population has most likely found ways to cope with the lack of indoor plumbing, it is the ignorant tourists who actually make contact with their own excrement. The assault targets the intimacy of the body, as the feces attacks the ankle of tourists that were, just a page earlier, described in all their fleshy exuberance as "unattractive, fat, pastrylike-fleshed" (13). Brian Larkin defines infrastructures as "matter that enables the movement of other matter," including excrement and the biomass of tourists. But infrastructures, Larkin notes, are not only things, but also "the relation between things" (2013, 329). And because they are essentially relational, infrastructures not only enable divisions between the haves and the have-nots, they also promiscuously and unpredictably connect different human, nonhuman, and (in the case of excrement) liminal entities. What appears in the passage as "locally grown food" turns out to be the output of a global delivery system: "it's better that you don't know that most of what you are eating came off a plane from Miami," where it "came from a place like Antigua first, where it was grown dirt-cheap, went to Miami, and came back" (Kincaid 2000, 14). Infrastructure, Kincaid shows, puts everything in relation, even if those relations, or consciousness of those relations, are deeply unwanted.

What, then, is the politics of this passage? In his reading of it, Bruce Robbins (2017) refers to the "you" as "an aggressive 'you'" (38), because it punishes the addressee for the global inequality the passage so bitingly exposes. Robbins (2007) feels the addressee—the tourist, but also privileged audiences like us, and also a professionally successful émigrée like Kincaid herself—does not deserve such punishment: "It cannot be our goal to have all those who benefit from indoor plumbing, wherever mobile or stationary, spend their time thinking about where the contents of their lavatories go after they flush" (33). Robbins is right that this cannot be our goal (and I am not sure anyone ever claimed obsessing over feces was a political *goal*), but he also misreads Kincaid. For Kincaid, such awareness, such compelled visibility, is not a goal so much as a *strategy*. The passage's repetition of "you must not wonder" amounts to telling the reader not to think about an elephant: it makes the lack of a proper sewage-disposal system impossible to unsee, and makes sure that the excrement cannot possibly become merely ambient again. What the passage denies its audience is ignorance and blindness. In that way, I think, it drives home the message that the (in)visibility of infrastructure is constructed and not merely given, and therefore open to change. Mundane, unremarkable infrastructures can be a deep aspiration, and, the passage insists, until access to clean water, adequate supply systems, and other key infrastructures is shared more widely, privileged audiences cannot feel entitled to it. If we want to simply rely on our utilities without having to obsess over them, these utilities need to be shared more broadly. As long as they are not, there is always the uncomfortable threat that we may connect with the excess matter we thought we had disposed of.

Kincaid's passage encapsulates the contemporary reality that privileged constituencies transport their waste to the Global South, where it renders toxic the lifeworlds of the dispossessed. This turns the lifeworlds of these people into sacrifice zones condemned to irreparable environmental decay in the service of economic growth. Economist Giovanni Arrighi (1995) has referred to this process as "the unplugging of … 'redundant' communities and locales from the world supply system" (330). Yet in an Anthropocene world, such opportunistic outsourcing cannot fail to rebound, as, apart from obvious ethical problems, such environmental racism also affects the habitats of the food we eat and the places we travel to, and creates geopolitical tensions that render all life vulnerable. Brian Thill (2015) diagnoses contemporary life with an "Away-fantasy": the fanciful notion that we can conclusively abandon the stuff we produce and consume (109). Quoting Timothy Morton, he remarks that "we may have thought that the U-bend in the toilet was a convenient curvature of ontological space that took whatever we flush down it into a totally different dimension called *Away*," but we now know that everything remains on the surface—on "viscous surfaces from which nothing can be forcibly peeled" (50). For Thill, hoarders living among seemingly infinite piles of stuff may seem socially deviant, but they are not more aberrant than "those of us who do the seemingly normal and healthy thing of dumping our mountains of trash into unseen dumpsites" (110). When we talk about waste, *not seeing* is a matter of privilege that requires hard, and occasionally violent, effort.

The digital age exacerbates this toxic juncture of infrastructure and waste. The very rhetoric through which digital culture promotes itself uses notions such as "wireless," "the cloud," and "instant connectivity" to suggest that communication and value creation take place in a frictionless vacuum, as if they were not enabled by thoroughly material infrastructures of fiber optic cables, massive data servers, and energy provision. Digital culture thrives on what Sean Cubitt (2017) has called an illusory "separation of the dirty business of generation from the clean image of energy consumption" (27). The business of producing electronic devices is "dirty" in two senses: first, it is thoroughly material, and dependent on vast amounts of often unclean energy and hard-to-mine rare-earth metals, and second, it requires the hard (and often compelled) labor of a vast army of workers exposed to toxic materials. Indeed, digital culture is itself a massive source of dangerous waste: as Jussi Parikka (2015) notes, "media technologies from monitors to game cartridges are abandoned, forgotten, stashed away, but retain their toxic materiality that surpasses the usual time scales we are used to" (113). If traditional industries "at least indicated danger with their smoke stacks," in reality "the purified industries of computing [are] secretly just as dirty as the industrial ancestors" (111).

Infrastructure's own role in obscuring material reality further complicates the challenge of *reading* infrastructure. Not only are the invisibility or (hyper)visibility of infrastructure not ontological features but the result of material and socioeconomic processes; infrastructures can also be the means by which reality is rendered invisible. Because, unlike smoke stacks, fiber optic cables can be hidden under the sea and data servers located at remote locations, contemporary life may seem to have become as smooth as the slick surfaces of the smart phones that connect us to each other. This is nothing new: think of how the infrastructure of the slave ship, for instance, made drinking coffee and smoking tobacco strangely guilt-free pleasures in early modernity. Still, infrastructure's implication in its own concealment compounds the difficulty of reading the infrastructure assemblages in which we live. Caroline Levine (2010) has coined the notion "infrastructuralism" to name "the practice of attending closely to the jostling, colliding, and overlapping of social, cultural, technological, and natural forms" (65). For Levine, "[i]f we want to describe the basic infrastructure of a particular, local site, what we find is a vast variety of chaotically overlapping, repetitive social forms that extend from multiple pasts and replicate themselves, indefinitely, into unpredictable and distant futures" (96). Levine's emphasis on repetition, multiplicity, and large temporal scales (from multiple pasts to distant futures) points to a critical practice that is not so much focused on revelation and inversion, but is rather attentive to the complexity and variety of the constellations in which we encounter infrastructure. Shannon Mattern (2013b) has written about "infrastructural literacy": the capacity to visualize and comprehend the way infrastructure takes shape and impacts our lives. Even if making visible the invisible and raising awareness about imperceptible infrastructures do not, as Mattern underlines, amount to significant political action, such literacy has an undeniably political dimension. Infrastructures, as I argued, are fundamentally enmeshed with desires and with collective aspirations; they encode, in Brian Larkin's words, "the dreams of individuals and societies and are the vehicles whereby those fantasies are transmitted and made emotionally real" (Larkin 2013, 333). In this overdetermined context, literature can intervene and leverage infrastructure for imagining an Anthropocene politics.

Off the Road/On the Grid: Literature Dreaming of Infrastructure

Los Angeles holds a special place in the Anthropocene imagination. With Hollywood, it is home to the world's dream factory that has, especially since the start of the millennium, steadily been producing images of environmental doom. Situated on the San Andreas Fault, it is a site of extreme geological vulnerability. This vulnerability is exacerbated by climate change-induced threats of coastal flooding and desertification. As Karen Tei Yamashita's 1997 novel *Tropic of Orange* notes, "climatic change in L.A. was different from other places. It had less perhaps to do with weather and more to do with disaster" (2017, 34). *Tropic of Orange* carefully resists the lure of disaster, and instead explores the multiple vulnerabilities besetting a globalized and overheating world by zooming in on, precisely, infrastructure. At the heart of the novel is the L.A. freeway system, a vast and iconic road network constructed in the middle of the twentieth century by the massive displacement of earth matter and local communities. Since then, it has continued to foster fantasies of fossil-fueled automobility (one critic calls it "the ultimate anti-'nature' locale" [Sze 2000, 34]). In *Tropic of Orange*, traffic literally comes to a standstill and makes room for different forms of mobility and different uses for infrastructure. If traditional disaster movies make infrastructure visible by blowing it up on screen (Rubenstein et al. 2015, 575), *Tropic of Orange* does something less ostentatious: it *suspends* infrastructure's customary operations to envision different forms of connectedness.

Tropic of Orange consists of 49 chapters and follows the intersecting trajectories of seven characters over seven consecutive days. The title of the table that precedes the novel, "HyperContexts," situates the novel in the context of the rise of the internet as the key aspirational infrastructure of the 1990s. *Tropic of Orange*'s exuberant postmodern style—mixing pastiches of popular culture, television formats, migrant fiction, and L.A. noir—reflects its alliance to the 1990s obsession with global connectedness. Yet the novel is savvy enough to underline the violence and inequality besetting global integration. Written shortly after the NAFTA (figured in the novel as a fairly ridiculous wrestler called SUPERNAFTA), which turned Mexico into a vast sweatshop catering to a rapidly deindustrializing USA, the novel is attentive to the relations of exploitation that organize L.A.'s relation to the South of the US–Mexican border and to its own undocumented workforce. Two of the main characters are first-generation immigrants who, "[e]ver since [they've] been here, never stopped working. Always working" (71). By opening the novel with the story of Rafaela, who is housekeeping in Mexico and lives away from her partner, the novel signals its commitment to making visible this invisible workforce. And if most of the other characters are of mixed descent—like Yamashita herself, who is Japanese American and lived in Brazil for a long time—the novel is no celebration of multiculturalism. NAFTA opens borders for capital, but it also closes them for the global poor. As one of the novel's subplots intimates, it also generates the conditions for illegal organ trade across the border. *Tropic of Orange* is more interested in the realities and complexities of connectedness than in providing empathetic connections to the different characters. As Rachel Greenwald Smith (2015) notes, the novel "establishes the primacy of relation over identification without sentimentalizing the commodified forms of connectivity that were beginning to solidify in the late 1990s: the commercialized Web, shallow senses of 'globalization,' and cable news-generated information" (22).

The importance of relation is reflected in the novel's thematic focus on infrastructure. The different stories coalesce around a vast traffic jam that blocks the L.A. freeway system: it is "the greatest traffic jam the world had ever seen" (178). This blockage converts the freeway into an improvised living space for the homeless who take up residence in the abandoned cars, making it "the greatest jam session the world had ever known" (177). The stability of infrastructure, it seems, enables it to outlast its original use (automobility) and become a site for spontaneous modes of sociality and mobility: "Amazing thing was everybody in L.A. was walking" (187). The sudden immobility of what the novel calls "the truck beast, whose purpose was to transport the great products of civilization" creates a vacuum that is immediately filled with life "reorganizing itself in predictable and unpredictable ways" (104–105). The emphasis is on the improvisational aspect of this innovation: the community, we read, is "quite a mess" (106), not a utopia. What is key is that infrastructure that used to sustain a particularly toxic way of life and mobility is repurposed for different modes of imaginative and social transport. If a different world is possible beyond the deadlock of fossil fuel modernity, the novel seems to suggest, it will still rely on infrastructural support.

The novel's repurposing of infrastructure goes together with its critical revision of globalization. This revision tends toward what we earlier, in the second chapter, identified as a properly *planetary* outlook. The main magical element in the novel is a peculiar orange that grows at the latitude of the Tropic of Cancer, the most northern line where the sun can be directly overhead, and an imagined border between the Global North and South. The emergence of this unseasonal "orange that should not have been" is immediately linked to global warming (14). The orange is a node in the transnational network the novel presents, as the production of oranges connects different parts of the American hemisphere, participates in global trade relations, and in a strange way also triggers the traffic jam at the heart of the novel, which is caused by an orange spiked with drugs. The special orange has the particular capacity to make visible the division between north and south: it is accompanied by "a very thin but distinct shadow stretched in a perfectly straight line along the dirt and sand" (15). By making visible an imagined border, the orange also makes it possible to manipulate that border, and alter the make-up of the globalized world. This happens through the quasi-mythical character of Arcangel, who, embodying the violent legacies of empire, colonialism, and slavery, "functions as the testimony of the indigenous, the displaced, the exterminated, the poor, and the workers" (Lee 2007, 510). Accompanied by a caravan of Mexican immigrants, Arcangel brings the orange across the border and drags the Tropic of Cancer with it. This "slow northward creep" (Rody 2009, 33) disorganizes the maps through which globalization tries to master the earth, as "distances [are] skewed and the streets [aren't] parallel" (Yamashita 2017, 194). It destabilizes the temporal and spatial coordinates on which globalization depends and restores a properly *planetary* dimension: temporal experience is marked by "an eerie liquid elasticity" (103), by "an uncanny sense of the elasticity of the moment, of time and space" (107) in which "[t]he world teeter-tottered" (119). The novel strongly suggests that the infrastructures of trade and transport can outlast their participation in capitalist globalization, and can sustain a properly planetary mode of organization.

Tropic of Orange's infrastructural literacy pedagogy is mainly conducted by Manzanar Murakami, a Japanese American character named after the Manzanar Concentration Camp for Japanese minorities where he was born during the Second World War. Formerly a surgeon, Murakami now performs as a conductor for L.A. traffic: he is "a sooty homeless man on an overpass" conducting "the greatest orchestra on Earth" (33, 35). Manzanar's body registers the sensory reverberations of the city—he senses "the time of day through his feet, through the vibration rumbling through the cement and steel"—and manages to convert them into a different affective complex: "a controlled reverie of rhythmic cadence and repeated melody" (32). Manzanar's sensitivity to the city and its community is emphatically linked to his attunement to the infrastructural layers that constitute this location. He sees layers of maps which "began within the very geology of the land, the artesian rivers running beneath the surface," which are later complemented by "the man-made grid of civil utilities: Southern California pipelines of natural gas; the unnatural waterways of the Los Angeles Department of Water and Power, and the great dank tunnels of sewage" (52). Manzanar's composite vision brings together "the prehistoric grid of plant and fauna and human behavior" with "the historic grid of land usage and property" and "the great overlays of transport—sidewalks, bicycle paths, roads, freeways, systems of transit both ground and air" (52). Manzanar's choreographic intervention, which clearly serves as a model for the novel's own imaginative operation of bringing intersecting lives in concert with one another, shows how infrastructure can provide the basis on which a more sustainable relation between human and nonhuman agents can be entertained. The novel notes that "the initial grid" of the city, "the railroads and the harbors and the aquaduct," was "built by migrant and immigrant labor" (203). *Tropic of Orange* shows how literature can play a vital role in anticipating "a new kind of grid" (203): a form of community built on something as banal and basic as shared infrastructure.

Energizing Infrastructure

One consequence of the massive traffic jam in *Tropic of Orange* is a radical decrease in the demand for energy. If initially, "[s]peculations arose as to how much fuel was required to keep an idling engine idling. How much rev to keep a battery alive," energy soon proves to be "a minor concern" (Yamashita 2017, 145). Changing our forms of life, the novel suggests, can trigger an overhaul of reigning energy regimes. Etymologically, the term "energy" derives from the Greek term *energeia*, which, in contrast to the notion of *dynamis* that names a mere potentiality, indicates a specific goal-oriented action, a form of actualization, an effective setting-to-work (Pinkus 2016, 25). During the traffic jam, progress has literally come to a halt, and this kind of *energeia* is disabled. Yet when the passage underlines that energy is still needed "to keep an idling engine idling," it links energy to the preservation of a mere potentiality for action. This points to a modern transformation in our understanding of energy: as Vivasvan Soni (2017) explains, modern energy has severed its link to specific goals and purposes, and has come to be conceived as a reserve, as an "infinite potentiality" that can be mobilized for different ends (134). Not directly linked to an immanent goal, energy, in the modern sense, names the self-evident assumption that there is an infinite quantity of resources to fuel human culture. Modern life, in other words, relies on the idea, which we now know to be false, that there is an endless supply of fossil fuel to keep us going.

One reason for the privileged relation between energy and infrastructure is the sheer vastness of the infrastructures that fuel modern life. "Global oil infrastructures" make up a "massive infrastructural network" including "5 million producing wells," "more than 2 million kilometers of surface pipeline," "6,000 fixed oil platforms and 635 offshore drilling rigs; and more than 4,000 oil tankers moving 2.42 billion tons of oil and oil products every year" (Appel et al. 2018, 19–20). The sheer size of oil infrastructure is a key factor in its perpetuation, and explains the "grave inertia in the face of the needed retrofit or conversion to other fuel sources" (20). This combination of ubiquity and immobility not only asserts itself in material terms, but also in the domain of culture: because the fantasies of mobility and comfort that oil sustains are "emblematic of all that is dynamic and disastrous in advanced capitalism" (Hitchcock 2010, 81), oil's very pervasiveness blocks access to alternative imaginaries; as Peter Hitchcock notes: "It is oil's saturation of the infrastructure of modernity that paradoxically has placed a significant bar on its cultural representation," and therefore also on the capacity to mount a cultural challenge to its power (81). Energy, and oil more specifically, are "everywhere and nowhere, indispensable yet largely unapprehended, not so much invisible as unseen" (Wenzel 2017, 11). And because energy and oil are so rarely seen and problematized, this state of affairs maintains what Imre Szeman (2011) has called a "fiction of surplus": "the belief that there will always be plenty of energy to go around," and that lack of energy or the destructive effects of energy consumption will never dramatically affect our ways of life (324). Energy and oil, that is to say, sustain dominant but destructive forms of life, and they resist becoming the occasion for an emergency.

The emergent field of the energy humanities situates itself in this imaginative and material impasse to develop a pedagogy for "[l]earning to read for energy" (Wenzel 2017, 13). This field takes it as given that our dependence on oil is unsustainable and that an energy transition is unavoidable. If we want to manage, rather than undergo, this transition, we need a sense of emergency. For the transition to be intentional rather than compelled, we need to take into account "where we sit historically, where we find ourselves in terms of our infrastructural dependencies and our affective and erotic attachments to the fossil economy" (Petrocultures 2016, 19). Patricia Yaeger (2011) has called for the need to bring our "energy unconscious" to cultural awareness in order to make our emotive attachments available for debate (306). Yaeger gives the example of Jack Kerouac's novel *On the Road*, a classic paean to American freedom and mobility. What happens, Yaeger wonders, when we ask "how often … Dean Moriarty and Sal Paradise stop for gas?" Reading for the text's (and the culture's) "energy unconscious" reveals that Kerouac's protagonists are "gasoholics": "Oil dependency created their world" (306). By making these unseen affective investments visible, the energy humanities help dismantle the "fiction of surplus" and allow energy infrastructures to emerge as an emergency in "the petromodern present" (Wenzel 2017, 4).

Imre Szeman, one of the leading thinkers in the energy humanities, recognizes three dominant discourses that discourage us from confronting this emergency. "Strategic realism" reduces the unsustainability of our dependence on oil to a question of strategic manipulation and geopolitical maneuvering that allows particular nations "to keep [their] economies floating in oil" (2019, 98). "Techno-utopianism" recognizes the harmfulness and untenability of our oil dependency, but trusts that science and technology will mitigate, or even postpone, the end of oil. Both strategic realism and techno-utopianism believe geopolitical or technological fixes will make altering our socioeconomic forms of life unnecessary; they obviate the need to consider energy as an emergency. As Szeman writes, they "remain committed to capitalism and treat the future as one in which change has to occur (new geopolitical realignments, innovations in energy use) if change at other levels is to be deferred (fundamental social and political changes)" (104). As the Petrocultures Research Group (2016) remarks, "the optimism usually attached to renewables is that *they make the world made by oil possible after oil*" (68). A third and decidedly less pro-capitalist discourse does recognize the need for social change, but fails to think of a realistic way to bring it about: "apocalyptic environmentalism" paints the post-oil future in disastrous terms as "a hell on Earth, obscured by a choking carbon dioxide smog" (Szeman 2019, 104). The only option this alarmist perspective sees is a transition through catastrophe. Cumulatively, the different forms of denial, mania, and disavowal that these three perspectives exemplify point to affective and imaginative blockages that Stephanie LeMenager (2014) has influentially defined as "petromelancholia": an "unresolved grieving" of the fiction of energy surplus and of the ways of life it made possible (16). Petromelancholia consists in an inability to let go of the "happier affects" of a life made convenient by the availability of cheap energy (102). It inhibits concerted efforts to imagine and construct more sustainable forms of life, although LeMenager, like other prominent thinkers in the energy humanities, believes that this melancholia can potentially be channeled into activism.

This brief discussion of the energy humanities brings together this chapter's focus on the difficulty of experiencing the petromodern present as an emergency with the topic of the final chapter: our altered relation to the future and the worry that there might be no human future left. One novel that is as obsessed with infrastructure as with the future is Ben Lerner's 2014 New York novel *10:04*. The novel is narrated by a writer who closely resembles the novel's author, and who is also called Ben, as he negotiates the challenges of reproduction, creative inspiration, and climate change. The novel is shot through with liberal guilt over the narrator's involvement in the destruction of the planet. As Ben De Bruyn has shown, the novel's reflections on more sustainable ways of life are often conspicuously linked to infrastructure: the novel features uplifting encounters on the subway, on train platforms, and on the bus, to the point that "public utilities," writes De Bruyn (2017), "seem to function like the positive counterpart to capitalism's destructive network of commodities" (966). These moments of community are triggered by what the novel refers to as "the vulnerable grid" (Lerner 2014, 4): the availability of infrastructures that are both necessary and precarious. An epiphany occurring fairly early in the novel connects an awareness of infrastructural dependency to vulnerability and a proleptic imagining of a more collective future:

> I was aware of the delicacy of the bridges and tunnels spanning it, and of the traffic through those arteries, as though some cortical organization now allowed me to take the infrastructure personally, a proprioceptive flicker in advance of the communal body.
>
> (28)

The metaphorical slippage between traffic and organic imagery resembles nothing so much as ex-surgeon Manzanar's choreographies in *Tropic of Orange*, which also serve to connect the personal to the collective. In *10:04*, also, this association is most intense when infrastructure has stopped doing its regular work, as if the "proprioceptive flicker" of future community is only visible during a blackout. The novel ends during the 2012 blackout that hit Manhattan at the time of Hurricane Sandy, which leaves room for infrastructure to be energized by something else than fossil fuels: "A steady current of people attired in the usual costumes was entering the walkway onto the bridge and there was a strange energy crackling among us … What I mean is that our faceless presences were flickering" (238). It is remarkable how quiet and non-apocalyptic this anticipation of the future is, as if radical aspirations are kept in check by an awareness of our continued reliance on some sort of infrastructure, some sort of energy. The novel's (almost) final intimation of a kinder future is a case in point: "we will catch the B63 and take it up Atlantic. After a few stops, I will stand and offer my seat to an elderly woman with two large houseplants in black plastic bags … Everything will be as it had been" (239). The novel's motto, lifted from Walter Benjamin, underlines the quiet messianism that characterizes the novel's reflections on infrastructure and politics: when the messiah comes, it will not be to radically alter or abolish the world, but to change everything by leaving most things the same—"Everything will be as it is now, just a little different."

It would be a mistake to read this is a self-defeating form of petromelancholic quietism: rather, the novel's hyper-self-conscious and incessantly worrying narrator *does* paint the present as a site of emergency, but he refuses to waste his energies on a facile apocalypticism—an attitude that, as we have seen, threatens to hinder the possibility of an intentional transition and to force us into a chaotic and spasmodic transformation. By recovering the future as an open-ended site of hope for the "infrastructure species," *10:04*'s textual operations come in line with what the Petrocultures Research Group sees as the key role of the humanities in managing the energy transition: the humanities cannot foretell the future, but they can "open us to a thoughtful and responsible composure towards [the future's] uncertainties and possibilities" (2016, 24). The humanities can "teach us not to fear difference when we can no longer retreat into the same" (24). *10:04*'s commitment to connecting the present emergency to a residual hope in the future exemplifies this mandate. At the same time, the fact that that commitment takes shape as a bare "flicker" testifies to the difficulty of this challenge. As the last chapter will show, this difficulty often converts the future into a site of anxiety rather than hope.

6 Residues

The Poetics of Extinction, or, Beach Reading

Human life in the Anthropocene is lived in the shadow of extinction. Over the vast span of geological time, life morphs into nonlife, and biological life forms transform into geoforms. The basic entanglement of human and nonhuman lives means that human life is not exempt from this process. It is not just that "starfish and sea urchin skeletons recompose into limestone" and that "stegosaurus recomposes into Brent crude" (Scranton 2018, 52): the more radical point is that human life participates in this geological cycle. When we confront patterns of seasonal change and individual life cycles, we can draw on a vast cultural repertoire of soothing images of regeneration and continuity; in the case of geological change, the truth of the contingency and fragility of our forms of life allows for no consolation or relief. One day, the earth will move on without us, or, to be more precise, with only a rocky residue of us (it will be one of the key points of this chapter that the difference between these two prospects is an important one). For Roy Scranton, the "motivating enigma" of the Anthropocene is the mismatch between our lived experience and our afterlife as stony remains: "the human as echinoderm, mortal flesh as immortal rock" (2018, 52).

One could argue that in the middle of the nineteenth century, Darwin's theory of evolution already made the threat of species extinction inescapable. Still, Darwin's *On the Origin of Species* (1859) and its dominant reception downplayed this sense of impending doom in at least two ways. First, Darwin's intervention emphasized the shared *beginning*s of human and nonhuman species: the fact that, as the very last words of the book have it, "from so simple a beginning endless forms most beautiful and most wonderful have been, and are being evolved" (2001, 490). The book's last paragraph conveys another central tenet of

Anthropocene thought, the entanglement of different life forms, as it invites the reader to "contemplate an entangled bank, clothed with many plants of many kinds, with birds singing on the bushes, with various insects flitting about, and with worms crawling through the damp earth" (489). Darwin's second book, *The Descent of Man* (1871), would make unmistakable the implication that such multispecies entanglements also extend to human life. Overall, the emphasis on the actuality and the continuity of the interweaving of different life forms shifted attention away from the inevitable finitude of biotic community. This shift was compounded by a second feature: the prevalence of notions such as "adaptation" and "fitness" in dominant interpretations of Darwinian evolution, which tend to flatter human life as a species that is somehow particularly adept at fighting the struggle for life and surviving what Darwin's last paragraph calls "the Extinction of less-improved forms" (490). Extinction, it seems, is a problem for other species. Uniquely among the species, human life has improved itself in a way that allows it to contemplate nature's "entangled bank" from a safe and scientific distance, and this somehow reconfirms its exceptional status within the multispecies assemblage Darwin envisions. As Claire Colebrook (2012) has remarked, this means that evolution was "thoroughly humanized, allowing 'man' to see all his technical extensions … as adaptive extensions of his organic and self-serving life."

The emphasis on entanglement and human distinction obscures that life forms not only change through struggle and competition, but also through random external events. One example is the so-called Cretaceous-Paleogene extinction event, situated some 66 million years ago, when the impact of a comet or asteroid devastated the earth system and led some three-quarters of the plant and animal species on earth to extinction. Yet extinction doesn't have to be that spectacular: as Joanna Zylinska (2018) notes, species are going extinct all the time in a prolonged process geologists call "background extinction," and "extinction is first and foremost a process rather than an event: it is an inextricable part of the natural selection that drives evolution" (52). "Extinction," Claire Colebrook (2018a) notes, "is as natural and inevitable as emergence" (150). Even if a focus on fitness and entanglement obscures the crucial role of extinction, major extinction events, such as that killing all dinosaurs 66 million years ago, are merely intermittent intensifications of ongoing processes that are constitutive of life.

There are at least three reasons for the current fascination with extinction. First, the extinction rate of species has recently intensified spectacularly in what Elizabeth Kolbert (2014) has analyzed as "the sixth extinction." This time, it is human actions rather than anonymous comets that are decimating global biodiversity. This means, second, that current concerns over biodiversity conservation and the protection of endangered species are framed by far-reaching uncertainties about the viability of the human forms of life that are devastating the planet. Institutions like capitalism, globalization, and democracy that have to a large extent defined Western modernity run up against their limitations in the face of rampant species extinction. As Ursula Heise (2016) has shown, environmental concerns over endangered species only "gain sociocultural traction to the extent that they become part of the stories that human communities tell about themselves" (5); more particularly, they resonate with doubts over the tenability and resilience of the forms of life that sustain us.

In a provocative short book entitled *Learning to Die in the Anthropocene*, Roy Scranton (2015) argues that preserving human life in an of age mass extinction compels us to take leave of our petrocapitalist forms of life: the challenge for "humanity to survive in the Anthropocene," he writes, is "to live with and through the end of our current civilization" (22). What is needed, Scranton (2018) notes, is "to let our current civilization die, to accept our mortality, and to practice humility" (8). Scranton's rhetoric of extinction and survival points to a third reason for the current craze with extinction: if Darwin's lesson of the entanglement of human and nonhuman lives left room for a redemptive belief in the exceptionality of human life, the Anthropocene closes off this intellectual and affective escape route, as we see that human life intensifies planetary processes of extinction and waste to the point that they also threaten our own species' survival. Human life, in other words, is not only an engine of extinction, but is also one of the "less improved forms" subject to extinction. A world without mammoths, passenger pigeons, and dodos, to name only three of the most charismatic extinct species that inhabit the Anthropocene imagination, is no longer a world displaying the triumph of human prowess, but a planet that will one day persist without humans.

Imagining an earth without humans is a popular feature of Anthropocene culture. We can think of Alan Weisman's popular science bestseller *The World without Us*, which pictures a world from which human life has suddenly vanished. Weisman's (2008) book describes in great detail how nature reclaims the planet in "revenge for our smug, mechanic superiority" (16). Weisman's posthumous story is told from the perspective of a confident "we," and is fortified by insights from a broad range of scientific and technological disciplines. Other instances of the "disanthropic" imagination are less self-assured, and obsess over the question of *how* human residues will be read and interpreted, if at all. If there are no more humans, or manmade disciplines like geology or history, to guide the interpretation of the imprints of human action, how (if at all) will nonhuman readers make sense of them? In his 1991 book *War in the Age of Intelligent Machines*, Manuel DeLanda envisions a future "robot historian" looking back on the history of technology. Such a robot historian would highlight different things than his human counterpart: it would, for instance, "hardly be bothered by the fact that it was a human who put the first motor together"; nor would it focus on the role of heroic individuals, "for these might be seen as mere catalysts for the self-assembly of war machines" (1991, 3). From this defamiliarizing perspective, human intentions don't count for much, as they are not even legible from the robot's optic. In his book of popular science *The Earth after Us* (2008), Jan Zalasiewicz (2009) imagines "extraterrestrial visitors from the galactic empire" who, millions of years in the future, attempt to reconstruct the legacy of our "major, intelligent yet transient civilization" (4, xiv). The result is an account of human life's geological agency that leaves many aspects of civilization we find self-evidently significant rigorously illegible. Often, projecting nonhuman readers of our human track record forces us to imagine a future in which human action will have been meaningless.

This anxiety over cosmic powerlessness and insignificance allows for a slightly different take on our obsession with extinction: it may have less to do with a fear to disappear without a trace than with the disconcerting realization that our traces *won't* disappear, but that they will no longer be ours to interpret. This lack of interpretive control resonates with the insight that the Anthropocene, in Bronislaw Szerszynski's (2012) words, witnesses human life's "becoming-mineral," its "incipient minerality," which transforms it from a meaning-making animal into a rock "to be contemplated by the geologist-to-come" (179–81). This alienating future perspective changes the way we inhabit the present, as, in Claire Colebrook's (2017) words, "we look at the earth—now—as if, in our future absence, we will be readable as having been" (6). When human life will have been recomposed into rock, this rock "will offer a 'reading' of a species' history," but that species will not have a say in its history's interpretation (Colebrook 2014, 24). Our record will be legible, but not by us, which is to say that it will be exposed to (mis)interpretation by others. The anticipation of our species' future as a mere residue puts paid to the illusion of human sovereignty and control.

The link between extinction and unruly meaning underlines the peculiar relevance of literature. Literature has long been troubled by the unpredictability of meaning and the inability to control the proliferation of signs. Think of P.B. Shelley's famous sonnet "Ozymandias," in which the boastful stature of an ancient "King of Kings" is reduced to a mere "colossal Wreck" consisting of "[t]wo vast and trunkless legs of stone" and "a shattered visage" abandoned in the desert. On the face of it, the poem expresses the impermanence of historical achievements. It recalls the famous ending of Michel Foucault's (2002) *The Order of Things*, which holds that "man is an invention of recent date" that will quietly disappear "like a face drawn in the sand at the edge of the sea" (422). Still, there is a world of difference between Ozymandias's "boundless and bare / … lone and level sands" and "the sand at the edge of the sea" where Foucault sees human life vanish. Ozymandias's visage is precisely *not* erased, but mercilessly survives into a context it had not anticipated, and in which its proud command to "[l]ook on my Works, ye Mighty, and despair" inspires less fear and respect than ridicule and pity. The point of the poem, in other words, is not that literary inscriptions are bound to disappear, but that they persist in contexts where their meanings and affects spin out of control.

The uncanny persistence of signs and the uncontrollable drift of language are key insights in literary criticism. However hard we try, meaning can never be fully guaranteed by the speaker's or the author's intention. Steven Knapp and Walter Benn Michaels's 1982 essay "Against Theory" is one of the most famous reflections on the relation between literary meaning and intention. The essay invites the reader to another stretch of sand, but one that is significantly different from both Foucault's site of erasure and Shelley's site of stubborn survival. Knapp and Michaels's shore is a thoroughly weird shore, where a nonhuman agent like the sea has the power to write (and here we find ourselves back on the coastline of Area X where we began this book). The essay evokes a beach where "a curious sequence of squiggles" spell out the first stanza of William Wordsworth's "A Slumber Did My Spirit Seal." While it would be possible to see this as a case of intentionless meaning, as meaningful writing generated by pure accident, this becomes substantially harder when a wave washes up and recedes, leaving in its wake the second stanza of Wordsworth's poem. For Knapp and Michaels (1982), the reader is now left with two options: either she accepts that these are "nonintentional effects of mechanical processes (erosion, percolation, etc.)," in which case this is not poetry, not language, but merely squiggles resembling words, or she posits an author (the living sea? The ghost of the poet haunting nature? a playful geoengineer?), in which case it does make sense to see the signs as an intentional act of expression (728–29).

Knapp and Michaels's official point is that the very idea of intentionless meaning does not hold water: as soon as we try to interpret the squiggles as meaningful marks, they submit, we cannot avoid positing the intention of an author. Yet in an Anthropocene context, their thought experiment acquires different resonances (although the point of their intervention is that such unintended resonances are strictly irrelevant!). When things acquire a form of agency, when matter is endowed with a sense of vibrancy, and when subjectivity, in Bruno Latour's (2017) words, "means *sharing* agency with other subjects that have also lost their autonomy" (62), erosion, percolation, and expression are no longer neatly distinct occurrences, and it becomes untenable to see the former as meaningless and the latter as intrinsically significant. In the context of climate change, for instance, the difference between intentionally harmful human actions and the unintended fallout of particular decisions becomes strictly irrelevant, as it is the effects, not the causes, that make up the reality of a climate changed world. In this context, sociologist Ulrich Beck (2015) has characterized the Anthropocene as "the age of side effects," in which unintended consequences are as significant as intended goals (78). Anthropologist Joseph Masco (2015) calls the Anthropocene "the age of fallout," in which the delayed destructive impact programmed into industrial and military processes manifests itself.

If we return from Knapp and Michaels's beach to P.B. Shelley's desert sand, we can see that Ozymandias's impotent boast is emblematic of literature's intimate concern with the impossibility of controlling the drift of meaning. Ozymandias's words are, in Shelley's poem, still recognized as meaningful language by the "traveller from an antique land" whom the poem's speaker encounters, even if their meaning has acquired unanticipated ironies and overtones, and even if we have to trust the traveler's powers of translation. It is less likely that alien geologists and robot historians in a future without humans will be able to appreciate the difference between stone carvings and accidental scratches. It is even less likely that they will care. The crisis that surfaces in the Anthropocene occupation with species extinction, then, has as much to do with a fear of powerlessness as with a realization of our mortality.

The rest of this chapter explores different affective and imaginative dimensions of that crisis. In the next section, I focus on what is no doubt the most popular literary form that engages the fragility of our current forms of life: post-apocalyptic (or post-catastrophe) fiction. Works like Margaret Atwood's *MaddAddam Trilogy*, Cormac McCarthy's *The Road*, and David Mitchell's *Cloud Atlas* are among the most studied contemporary texts dealing with environmental crisis. Through a discussion of Emily St. John Mandel's *Station Eleven*, another critical favorite, I argue that the genre is often marked by a conspicuous *denial* of extinction, as these texts' reluctance to imagine an alternative to current forms of life tends to turn them into exercises in continuity rather than into confrontations with rupture. As Theodore Martin (2017) has noted, post-apocalyptic novels are often "desperate for the mere continuation of the present" and tend to be "experiment[s] in monotony" (161–62). More enabling engagements with extinction, I argue in the chapter's third section, require a minimal *detachment* from current forms of life: an acceptance of the inevitability of extinction, and a readiness to, in Roy Scranton's words, "learn to die in the Anthropocene," or, in the terms of queer theorist Lee Edelman, interrupt self-perpetuating cycles of "reproductive futurism." Facing extinction paradoxically relies on a willingness to entertain the possibility that the future will be significantly different from the present.

Detachment from or disappointment with existing forms of life at times flip over into a studied *indifference* to human extinction. The fourth section of this chapter surveys a number of thought experiments by philosophers Nick Bostrom, Ray Brassier, Jean-François Lyotard, and Quentin Meillassoux that resituate human life in more capacious spatiotemporal orders, even while refusing to become sentimental about the cosmic downgrading of human existence. Such thought experiments have distinct literary qualities, and they helpfully destabilize the idea that the future can only be imagined as ever more of the same. At the same time, the "sado-dispassionate" character of this kind of "disanthropic" imagination, as Greg Garrard (2012, 44–45) has called it, threatens to shade into a more aggressively *misanthropic* program that thrives on the superfluity of the majority of human lives—or, in the case of philosopher David Benatar's anti-natalism, of the whole species. The upshot of such extinction fantasies is very often what Claire Colebrook (2018a) has called "species-bifurcation, with some humans commandeering and squandering the few remaining resources while enslaving the majority of barely-living humans" (151). I turn to Peter Frase's analysis of "exterminism," or the large-scale elimination of people without economic value, and William Gibson's novel *The Peripheral*, which I already discussed in the introduction and which imagines a future bifurcated species, to show how literary and critical thought can expose the sinister politics of toxic forms of extinction discourse. Denial, detachment, indifference, misanthropy: these four postures present so many ways of engaging the motivating enigma of the Anthropocene: the paradoxical relation between "mortal flesh" and "immortal rock," between human significance and nonhuman residue (Scranton 2018, 52).

Denial: Apocalypse against Extinction

The apocalyptic intimation that the world as we know it is about to come to an end is nothing new. The notion is customarily traced back to the ancient Iranian prophet Zarathustra (or Zoroaster), who bequeathed "a sense of urgency about the demise of the world" to Jewish and Christian traditions (Garrard 2004, 85). This intimation of destruction did not disappear when secular modernity displaced a theological framework, and intimations of the destruction of the world have become a staple of literary fiction, rather than only theological speculation. For Christian thought, apocalyptic anticipation brings home the vanity of earthly existence, which can be compensated by heavenly rewards; in modern life, the figure of apocalypse signals the perceived fragility of secular institutions: democracy, individualism, capitalism, technology. As Heather Hicks (2016) writes, "with the emergence of modernity in the eighteenth century, apocalypse shifted from its origins as the story of the annihilation of a sinful human world to become, in novel form, the story of the collapse of modernity itself" (2). That the institutions of modernity itself are at stake explains the affinity between apocalyptic musings and the Anthropocene: the Anthropocene is, among other things, a name for the exhaustion of modern forms of life. It also explains that recent apocalyptic fiction often takes recourse to a decidedly pre-modern world of extreme brutality and precarity, in which resources are depleted, technology has become defunct, and infrastructures are demolished. If science fiction typically offers a technologically enhanced version of the present, apocalyptic fiction routinely adopts the opposite mode of operation: it "doesn't intensify the present moment, it contradicts it" (Harbach 2008). Anthropocene apocalypse, then, is as much *post*-apocalyptic (as it imagines the aftermath rather than the destructive event itself) as it is *pre*-modern (as it obsesses over the demise of modern achievements).

Given such dreariness, how can we explain the long-lasting appeal of the apocalyptic imagination? One reason is that apocalypse is not only a matter of destruction, it is also a moment of revelation. The Greek *apokaluptein* means "to uncover, to reveal," and James Berger (1999) notes that an apocalyptic event must "in its destructive moment clarify and illuminate the true nature of what has been brought to an end" (5). Apocalyptic fiction, in other words, rewards us with an analytical insight into the world as it exists. It does so, moreover, without really threatening its audiences: consuming post-apocalyptic fiction is a way of courting near-extinction rather than confronting the fact of a coming collapse. In this respect, the appeal of imagined apocalypse is very similar to that of the romantic sublime: as in the encounter with cataracts, mountains, and other overwhelming natural forces, the terror threatening human survival becomes pleasurable because it is faced from a safe distance or, in the case of post-apocalyptic stories, through the frame of fiction.

This relative innocuousness of the post-apocalyptic mode accounts for the broad range of affects it adopts in contemporary fiction: the hectic weirdness of Margaret Atwood's *MaddAddam* trilogy, the relentless horror of Cormac McCarthy's *The Road*, the subdued suburban melodrama of Tom Perrotta's *The Leftovers*, the hardboiled cool of Paolo Bacigalupi's *The Windup Girl* and *The Water Knife*, or the ecstatic abandon of the central section in David Mitchell's *Cloud Atlas*. Such affective variation is a feature of the longer tradition of apocalyptic fiction: if Mary Shelley's 1826 classic *The Last Man* evokes the demise of modern civilization in anguished terms, later instances of the genre like John Wyndham's *The Day of the Triffids* from 1951 belong to what has been called "the 'cosy catastrophe' subgenre of apocalyptic fiction," in which "global disaster is survived by a typically prosperous remnant that adapts, with reasonable aptitude and plenty of common sense, to the new conditions of a post-collapse world" (Tate 2017, 8). Apocalyptic fiction affords an affective response that does not preclude coziness or consolation.

There is by now a long catalogue of names for varieties of apocalyptic narrative: not only post-apocalypse, but also neo-apocalypse, crypto-apocalypse, ana-apocalypse, and others (Hicks 2016, 6). This terminological instability reflects the paradoxical temporality of apocalypse: the fact, that is, that the end can only be narrated if it is survived. The very existence of apocalyptic stories proves that the supposed moment of ultimate disintegration is not as fatal as feared. In apocalypse, then, the ending "both does and does not take place" (Berger 1999, xii). By its very nature, a post-apocalyptic world "consoles the reader with the impossibility of remainderless destruction, with the apocalyptic triumph" (Dillon 2007, 377). The genre, in other words, is as much a fantasy of survival as a reckoning with extinction.

Emily St. John Mandel's 2014 novel *Station Eleven* exemplifies this mixture of anxiety and fantasy. The novel is organized around an epidemic that wipes out 99.6 percent of the world population, and it focuses on the surprising resilience of small sets of survivors. One such group settles in a Canadian airport terminal and manages to organize a tolerant multicultural community in what used to be a supermodern "non-place" (Augé 1995). The survivors even start a "Museum of Civilization," which collects defunct objects from the pre-catastrophe world: iPhones, Nintendo consoles, stiletto heels, a snow globe (Mandel 2014, 255, 258). Significantly, these objects' uselessness, in the post-pandemic world, makes them more beautiful: the characters "had always been fond of beautiful objects, and in [their] present state of mind, all objects were beautiful" (255). Beauty, it seems, depends on scarcity and obsolescence: the realization that things can go defunct makes the survivors appreciate things like electronic gadgets, airplanes, and simple commodities. This is a key operation of post-apocalyptic fiction: it removes objects from their habitual functions and invites us to appreciate and desire their fresh splendor, as if they were not part of the boredom and waste of everyday life. In this way, these fictions suggest that it is the overabundance and excess of contemporary life that robs the world of its beauty. As we will see, this at times comes precariously close to a disavowed desire to get rid of this surplus and to inhabit a less populous planet.

That *Station Eleven*'s post-pandemic setting serves to strengthen our attachment to the achievements of the modern world is also clear in the second privileged group of survivors. "The Travelling Symphony" is a group of actors traveling around the devastated world and performing music and Shakespeare plays. Shakespeare stands in for the best that Western culture has produced, a highlight that the Symphony take it upon themselves to keep alive after the catastrophe. Shakespeare, that is, is the question to which the novel's refrain gives the answer: "*Because survival is not sufficient*" (58, 119). Mere survival would be a situation in which human life persists as a bare biological entity, as a residue bereft of meaning. Post-apocalyptic novels do not surrender to such meaninglessness. In McCarthy's *The Road*, for instance, the father and the son are surrounded by cannibalism, violence, and destruction, and they tell themselves they "carry the fire": they adhere to a minimal code of conduct that somehow keeps them connected to an obsolete civilization.

The meaning of this code is never spelled out; it is as numinous and elusive as *Station Eleven*'s invocation of Shakespeare. Atwood's *MaddAddam* trilogy, for its part, is obsessed with the power of storytelling to transmit culture, in however warped a form. "Sloosha's Crossin'," the post-apocalyptic middle section of David Mitchell's *Cloud Atlas*, is situated in a future "Ha-Why" (Hawaii) where even the English language has deteriorated, and civilization has disappeared after what the novel refers to as "the Fall." Yet even here, hope in regeneration is kept alive by an "orison," an egg-shaped device that preserves sounds and images it projects as holograms. This orison transmits the testimony of a character from an earlier section, set in a dystopian, totalitarian future Korea where workers are cloned and fed the flesh of their kin. The fact that the survivors don't understand the testimony's words—as the novel has it, it "speaks in an' Old'un tongue what no'un alive und'stands nor never will, nay" (Mitchell 2004, 324)—is beside the point: it is the very fact of connectedness and continuity that safeguards the possibility of meaning. The threat that the residues of human culture will be abandoned to indifferent aliens is contained by imagining a diminished but decidedly human tribe of survivors.

The emphasis on small numbers and on continuity is a key feature of contemporary post-apocalyptic fiction. This sidesteps the real challenge of species extinction: the prospect that, just perhaps, Shakespeare's plays will survive but there will be no one left who knows or cares how to read or perform them. It also falls short of capturing the import of the Anthropocene in at least three further ways. First, the genre's reliance on a cataclysmic event—probably a nuclear incident in *The Road* and *Cloud Atlas*, an airborne disease in *MaddAddam* and *Station Eleven*—misrepresents the gradual processes of attrition and degradation that mark the Anthropocene's "violence of delayed destruction" (Nixon 2011, 2). Stephanie LeMenager (2017a) has argued that attending to this "slow violence" requires a mode of writing that lingers on such everyday realities, on an "everyday Anthropocene," rather than offer an "epochal discourse that capitalizes on the charisma of crisis" (225). For Joseph Masco (2015), the apocalyptic frame is a leftover from the Cold War effort to mobilize the public around nuclear threats, and it fits uneasily with the long, slow, and incremental rhythms of the Anthropocene. Post-apocalyptic fiction misrecognizes the differences between nuclear war (short and fast) and Anthropocene attrition (long and slow) and perpetuates a logic of militarization.

Many post-apocalyptic fictions do dwell on the realities of eroding lifeworlds and biodegradation: the meticulous description of the minutiae of everyday survival is one of the defining features of the genre (a feature that, as Heather Hicks has noted, makes Daniel Defoe's *Robinson Crusoe* a vital precursor for the genre [2016]). Yet this myopic focus on isolated objects and small groups also means that crucial aspects of contemporary life are *not* factored in (and this is a second limitation of the genre): Anthropocene life is not a life of auratic objects and neatly circumscribed groups, but a life of excess, waste, entanglement, and multiplicity. In her critical assessment of the genre, Ursula Heise (2015) notes that "[w]hat really counts is that the characters, in their break from the corruptions of the past, no longer have to deal with things like crowded cities, cumbersome democracies, and complex technologies." *Station Eleven*'s linkage of beauty with scarcity drives this point home: after the collapse, the world is rendered more lovely by the prospect that "[p]herhaps soon humanity would simply flicker out," which leaves room to enjoy "[t]he beauty of this world where almost everyone was gone" (148). A comparable strategy is at work in *The Road*. The novel's poetics of subtraction presents a biodegraded, exhausted, and barren planet in a clipped and hardboiled style, but it still promotes a small-scale continuation of modern life. The book's minimalism is not only a stylistic feature: the world it envisions is a pared-down one. As Chad Harbach (2008) notes, *The Road* "defends the world we know by giving us none of it." Meaning and pleasure depend on small numbers and are not available in an overcrowded world. Safeguarding meaning and pleasure then comes to depend on getting rid of surplus populations, a task these novels conveniently outsource to pandemics and disasters.

Dissatisfaction with excess and complexity points to the conservatism of many post-apocalyptic texts. By conveying a vision of human flourishing that is not just non-scalable, but that actively depends on exclusiveness, it offers what Heise (2015) calls a "weak cocktail of critique and complacency": critiquing social and environmental ills, these fictions often end up recommending a more rarified version of the status quo. The resistance to multiplicity situates contemporary post-apocalypse in the tradition of what Greg Garrard (2004, 93) has called "[t]he most influential forerunner to the modern environmental apocalypse," Thomas Malthus's 1798 *Essay on the Principle of Population*. Malthus argues that the limitations of the planet make continued societal progress impossible, as demographic growth will inevitably outpace agricultural production. For Malthus, perpetual conflict over scarce resources is unavoidable; as in contemporary post-apocalyptic texts, unchecked demographic expansion is the reason we can't have nice things. If Malthus's prognostics marked the industrial onset of the Anthropocene, Anne and Paul Ehrlich's *The Population Bomb* (1968) updated his intervention for the age of the Great Acceleration, prophesying that "hundreds of millions of people will starve to death" due to the exponential multiplication of human lives (Ehrlich 1971, xi). Malthus's intimation of an ever proliferating human biomass famously informed Darwin's notion of natural selection; just as Darwin's theory, as we saw, leaves room for the illusion that the human species will escape extinction, the Malthusianism of fictions of the post-apocalypse ends up sustaining a specious fantasy of survival.

The emphasis on punctual disasters and the resistance to multiplicity are not the only two elements of the genre that clash with Anthropocene realities; its unacknowledged provincialism is another one. The way of life that is at stake in these fictions is almost consistently a Western one: a comfortable, technologically advanced, and economically privileged form of life. It is not so much the species that is threatened, but the more rarified convenience of liberal affluent urbanity (Colebrook 2018b, 161). The wastelands in which the survivors in *Station Eleven*, *The Road*, and *Cloud Atlas* find themselves are only a diminishment for independent and self-sufficient modern subjects; for many people living in the Global South, they are often already a reality. And have been for a long time: as I explained in Chapter 4, current metropolitan anxieties merely show that the erosion of the lifeworld has finally caught up with privileged audiences, while "for the native people of the Americas, *the end of the world already happened*—five centuries ago" (Danowski and Viveiros De Castro 2017, 104). Post-apocalyptic fiction tends to project anxieties over a very particular form of life, and it blows up that concern to cosmic proportions, as if no other form of life were even imaginable. In the process, it obscures the possibility of what Claire Colebrook (2019) calls "a radical end that would not be an end for us, and that might generate another world" (264).

Detachment: The Afterlives of Extinction

My fairly critical discussion of post-apocalyptic fiction showed that its confrontation with species extinction is limited by its strong affective attachment to existing forms of life. The beach excursion that opened this chapter showed why detachment from these forms is hard: once the communities and institutions that give meaning to our actions and decisions disappear, we risk ending up as mere traces preserved in rock, exposed to the indifference of alien gazes. Detachment, in other words, requires a conviction that significance is also possible (if never guaranteed!) *beyond* the horizon of our customary forms of life. Philosopher Samuel Scheffler's 2012 Berkeley Tanner Lectures, published as *Death and the Afterlife*, make a philosophical case for the impossibility of detachment. For Scheffler (2016), the prospect of extinction robs our lives of all significance. Unlike the knowledge of our personal death, the certainty that there will be no humans after us would mean that many of the things that matter to us will stop doing so (15). Scheffler's argument has a crucial literary component, as he derives his conclusion from two thought experiments that convey "the prospect of the disappearance of the human race" (74): there is the doomsday scenario, which invites readers to imagine that the "earth would be completely destroyed thirty days after [their] death in a collision with a giant asteroid" (8), and there is a second scenario in which we envision sudden collective infertility (a scenario dramatized in, among others, the novel and film *Children of Men*). In the latter hypothetical situation, nobody gets hurt: we and our loved ones (who could potentially be directly affected by the asteroid in the first scenario) are very likely to die a peaceful death (39). What makes this scenario disturbing, then, is not my concern for myself or my loved ones, but a looming sense of the insignificance of all human projects in the face of the species' imminent demise.

For Scheffler, this proves that meaningfulness depends on the survival of the species. Scheffler draws uplifting conclusions from this insight: it points to a limit to individualism, as "the existence of the afterlife"—that is, other, often distant, people's lives—ostensibly "matters more to us than our own continued existence" (26). This sounds like good news: it makes it likely that people faced with the prospect of human extinction will do what they can to minimize the chances of planetary cataclysms, as everything they hold dear depends on the persistence of human life. Still, the earlier critique of post-apocalyptic fiction also applies here: what if such invocations of collapse are more concerned with the survival of a particular way of life than with that of the species? What will guarantee that a desire to continue a life of privilege will not bring people to fight constituencies that seem to threaten that entitlement? Scheffler writes that "[h]umanity itself as an ongoing, historical project provides the implicit frame of reference for most of our judgments about what matters" (60). He seems oblivious that "humanity" is, more often than not, a misnomer for a projection of Western ways of life—not least in official discourses of the Anthropocene. As in post-apocalyptic fiction, the specter of future destruction and infertility in Scheffler's argument serves to entrench a commitment to the status quo and forecloses the imagining of a different world.

Roy Scranton's *Learning to Die in the Anthropocene* draws different lessons from the imminence of extinction. For Scranton, the fact that our current forms of life have set us on an irreversible course to planetary desolation proves that these forms do not even deserve to be perpetuated. Our civilization has already stopped making sense: "Carbon-fueled capitalism is a zombie system," Scranton (2015) writes, and "this civilization is already dead" (23). His approach adopts the Buddhist insight that, just like the human individual, civilization is "impermanent, transient, and insubstantial" (Scranton 2018, 68), and translates it into a posture of humility rather than militancy. Scranton's detachment from the certainties of Western civilization makes room for the diversity so clearly lacking in Scheffler's book and in most post-apocalyptic fiction. He calls for diversity between cultures, but also between species: "We need to learn to see not just with Western eyes but with Islamic eyes and Inuit eyes, not just with human eyes but with golden-cheeked warbler eyes, coho salmon eyes, and polar bear eyes," or even with "the wild, barely articulate being of cloud and seas and rocks and trees and stars" (2018, 8). Scranton advocates a form of interdisciplinary imagining and description that, as we saw in Chapter 1, fits the mandate of the environmental humanities much more than the template of post-catastrophe fiction.

If Scheffler asks us to double down on the forms of life we inhabit, Scranton insist we should disengage from what he calls "the perpetual circuits of fear, aggression, crisis, and reaction that continually prod us to ever more intense levels of manic despair" (2015, 86). These cycles of excitement must, in a term Scranton borrows from the philosopher Peter Sloterdijk, be "interrupted" (rather than perpetuated) through thought, contemplation, and questioning (86, 108). This imperative to disrupt the reproduction of the same and recover an open-ended future resonates with work in the field of queer theory. Dedicated to the exploration of non-normative sexualities and identities, queer theory, in the words of José Esteban Muñoz, aims to "see and feel beyond the quagmire of the present"; "Queerness," Muñoz (2009) writes, "is essentially about the rejection of a here and now and an insistence on potentiality or concrete possibility for another world" (1). Queerness resists a future that simply reproduces the norms of the present. In his virulent polemic *No Future*, Lee Edelman relentlessly targets the figure of the child as the organizing principle of political life: the unconditional focus on the child, Edelman (2004) argues, constrains the political imagination as it cements "the consensus by which all politics confirms the absolute values of reproductive futurism" (3). Renouncing reproductive futurism and the heteronormativity it naturalizes requires a *refusal* to fight for the children. For Edelman, queerness consists in a stubbornly critical attitude that refuses to maintain any connection to a different future: "the queer comes to figure the bar to every realization of futurity, the resistance, internal to the social, to every social structure of form" (4).

Edelman's antisocial and antirelational version of queer theory has been criticized from perspectives that see more value in imagining novel communities and shared futures. Still, his emphasis on the unworking of reproductive patterns of the kind Scheffler takes for granted may be a vital step for intimating a sustainable future. Consider Ben Lerner's *10:04* again, a novel whose investment in infrastructure I discussed at the end of Chapter 5. One source of anxiety for the novel's narrator, apart from his concern over rising sea levels, is that his best friend Alex has engaged his services as a sperm donor, without really resolving the issue of his paternal involvement. As the novel writes its way toward a less inhibiting and more receptive relation to a future that is not constrained by financial, professional, social, and climatological doom, the narrator also comes to accept his paternity, even if the terms of the arrangement remain opaque. The novel ends with him accepting the child, but this prospect is not scripted by the dictates of reproductive futurism. The narrator and his friend walk to the doctor for a sonogram during the blackout caused by Hurricane Sandy. In the waiting room, they watch footage of the storm, "images of the swirling tentacular mass [spliced] with footage of it reaching landfall, of houses being swept away" (Lerner 2014, 232). In the doctor's cabinet, the images of the unborn child are not very different: "we see the image of the coming storm, its limbs moving in real time, the brain visible in its translucent skull" (233). The affinities between a climate changed world and the anxieties of reproduction signal a receptivity to a future that is more than a repetition of the present, even while it refuses the radical negativity called for by Edelman. *10:04* does not double down on the forms of life it evokes, but is, in Scranton's terms, a literary attempt to "interrupt" their compulsive reproduction and anticipate a different future in a climate changed world.

Such interruptions make imaginative room for other cultural forms of life and biological life forms. Extinction scholar Thom van Dooren (2014) emphasizes that the obligation "to hold open space in the world for other living beings" is a crucial ethical imperative in the age of the sixth extinction event (5). Richard McGuire's 2014 graphic novel (of sorts) *Here* is a good example of how literature detaches itself from present ways of life and makes imaginative space for different realities. *Here* consists of some 150 two-page drawings of one particular location at different moments through time. The first drawing situates us in a middle-class living room in 2014 before the book gives us the same room in 1957, 1942, and 2007. It then moves to 1623 to disrupt the domestic vibe and remind us of the longer history of the location. The historical transport between different drawings is further compounded by smaller panels opening up within the drawings that provide glimpses of different times, as when the 1986 living room is overlaid with two panels, one featuring a family discussion from 1972 and one offering a seemingly idyllic nature scene with an Indigenous American in the same place in 1352. McGuire's formal choices evoke a kind of porous temporality in which different scales and rhythms intersect—the layered temporality that, as I explained in Chapter 4, is so typical of our temporal apprehension of the Anthropocene. McGuire strictly adheres to the unity of place rather than, say, to the continuity of character, and this underlines *Here*'s resolute intent to move beyond anthropocentrism.

This ambition becomes even clearer when McGuire uses the affordances of the format to envision a world without humans. One panel is situated in the year 10,175 and features an empty field around a strange creature that one critic has described as "a marsupial of some sort, maybe the lovechild of a large possum and a small kangaroo" (Konstantinou 2015); another critic likens it to a "vaguely humanoid animal—almost something like the next evolution of H.G. Wells's Morlocks" (Canavan 2017). These divergent interpretations make clear that this is a creature that, even if it does not exist, still evokes enough cultural and biological reference points to offer some orientation, some familiarity; it is not radically other, only weirdly different. The relatively mild distinctiveness of the animal may be the point: it shares the pages of the book with children playing, people arguing or watching television, and other scenes of minor domestic drama. This future creature is not a sublime monster to be feared, but just part of the life of the planet. Some of the most exuberantly colorful plates in the book take us to the year 22,175, and feature dinosaur-like giants, nearly fluorescent birds, and lush, jungle-like plants. Again, this world without us is not presented as an occasion for anxiety or loss, but as a variant reality that can be welcomed once we detach ourselves from the grasp of current commitments.

My last literary example in this section is the "speculative fabulation" entitled "The Camille Stories" that ends theorist Donna Haraway's book *Staying with the Trouble* (2016, 134). This multispecies fantasy tells the story of five generations of human beings living between 2025 and 2425 who are born in symbiosis with "animal symbionts"; the first Camille, for instance, is born in symbiosis with a monarch butterfly. This is not so much reproductive futurism as what Haraway calls "exercis[ing] reproductive freedom with wild hope" (142). Haraway's fiction resolutely decenters human exceptionalism, without misanthrophically wishing for human extinction. It makes imaginative room for a future shared with other species. Camille, Haraway writes, is "one of the children of the compost who ripen in the earth to say no to the posthuman every time" (134). The Camilles operate in the ruins of a damaged planet and "work with human and nonhuman partners to heal these places, building networks, pathways, nodes, and webs of and for a newly habitable world" (137).

Haraway's project of "imagin[ing] flourishing with and for a renewed multispecies world" (134) not only makes *imaginative* room for nonhuman species, it also directly confronts the Malthusian issue of *physical space* (without outsourcing that solution to planetary catastrophe in the way post-apocalyptic fiction does). In Haraway's fabled future, human numbers modulate from a peak in 2100 (with ten billion) to "a stable level of 3 billion by 2400" (144). Because the "Communities of Compost" manage to convince the rest of the earth to adopt their life-sustaining and self-limiting mode of living, they provide necessary "breathing room" that, Haraway writes, opens up "possibilities for ongoingness for many threatened ways of living and dying for both human and nonhuman beings" (144). This concern with sustainable numbers addresses a crucial challenge for designing a livable future. The issue of population growth is mixed up in complex ways with many aspects of the Anthropocene: think of deforestation, soil degradation, excessive resource use, and geopolitical instability. Still, it has hardly become visible in its own right in environmental thinking: Timothy Clark (2016) talks about "population denialism" and the "insidious elusiveness" of overpopulation in most environmental discourse (10, 13). Voluntary self-limitation would have a vast beneficial impact, in the form, for instance, of a voluntary global One Child Policy. Such a policy, Alan Weisman has calculated, would mean a population of less than 6 billion by 2050 instead of the projected 9 billion, and a correspondingly smaller ecological footprint for our species

(Garrard 2012, 58). It would also, to recall Thom van Dooren's words, "hold open space" for other species to flourish.

Indifference: Cosmic Insignificance

Lerner's, McGuire's, and Haraway's fictions of regeneration face the threat of species extinction to envision a future beyond established ways of living, spending, and reproducing. The forms of conviviality they imagine ignore Scheffler's call to double down on these existing ways of life, while they also avoid the death-obsessed negativity of Edelman's flirtation with species suicide. They differ from post-apocalyptic fictions that situate significance in the continuation of the project of human civilization, and find meaning in different forms of life, even in forms of life shared with nonhumans (in the case of Haraway) or from which humans have been evacuated (in McGuire's posthumous bestiary). They emphatically remain committed to the possibility of significance, and mobilize their aesthetic strategies to avoid the threat of radical meaninglessness. This section surveys a number of thought experiments that take the detachment from extant forms of life one step further, and radically uncouple human life and significance. As we will see, once the shock value of such undisturbed declarations of senselessness wears off, their official indifference to human purpose all too easily shades into a misanthropic belief in the superfluity of some humans. All too often, the univocal declaration of species insignificance masks a sinister decision between valued lives and lives not worth living.

Jean-François Lyotard, one of the leading thinkers of the postmodern condition (the title of his famous 1979 book), begins the first chapter in his 1992 book *The Inhuman* with a thought experiment that positions human life in the shadow of planetary extinction. Lyotard reminds us that we know that the earth won't last forever, and that the sun will explode in 4.5 billion years. What changes when we contemplate life in the face of "the inevitable explosion to come" (Lyotard 1992, 9)? Lyotard is clear-eyed about the threat this preordained destruction of matter poses to human significance: "In comparison everything else seems insignificant" (9). There will be no future significance, because there will be no future readers to look for human meaning: "there'll be no one there to toll the death knell or hear it" (9), as there won't be "intelligent, sensitive, sentient earthlings to bear witness to it" (10). If post-apocalyptic fiction hinges on the consoling paradox that telling the story of the collapse proves that it has not been fatal, Lyotard's extinction event, in contrast, is radically unwitnessed. Lyotard wonders how human thought will continue to rage against the dying of the solar light, and how "the combined forces of nuclear physics, electronics, photonics and information science" are designing hardware that allows thought to continue without human bodies (14). This raises the specter of the "'separability' of intelligence" (16): the fantasy that reason can survive, even when human bodies cannot.

Lyotard denies that the binary thinking and data-processing of computers amount to genuinely meaningful thought. Thought does not take place in silicon micropressors, but requires desiring and suffering bodies: words and phrases only acquire meaning because "even inscribed on a page … they 'say' something other than what we 'meant' because they're older than the present intent, overloaded with possibilities of meaning—that is, connected with other words, phrases, shades of meaning, timbres" (18). This returns us to Shelley's desert, Wordsworth's beach, and Darwin's entangled bank: true meaning depends on the kind of semiotic drift and historical instability that literary thought has always cultivated.

Not everyone shares Lyotard's conviction that significance requires human hesitations and uncertainties. For a philosopher like Nick Bostrom, the "'separability' of intelligence" is an opportunity, not a curse, as the triumph of reason trumps the survival of the species. Bostrom runs the Future of Humanity Institute at the University of Oxford, and while it is obvious that this Institute is singularly unconcerned about the future of the *humanities*, it is more surprising that it is also weirdly complacent about the future of *humans*. What is at stake is human*ity*, understood as "*intellectual life as such*" (Colebrook 2018b, 162), and this humanity is not confined to human bodies. Bostrom aims to harness the affordances of technology and artificial intelligence to enhance human life beyond its natural limits to the point where humanity is "no longer unambiguously human by our current standards" (Bostrom 2014a, 3). This position often goes under the name of *trans*humanism, rather than that of (an often more technology-critical) *post*humanism. For Bostrom (2014b), the explosion that is bound to blow up our established forms of life is not solar catastrophe, but rather "an intelligence explosion": the emergence of a superintelligence that dwarfs humans' limited cognitive capacities, and that forces us to enhance human capacities, even if this involves surrendering our mortal bodies (4). This view is echoed by futurists like Hans Moravec and Ray Kurzweil, who strongly believe that marrying human consciousness to computer software will emancipate human intelligence from bodily constraints. Kurzweil (2005) has influentially dubbed the coming intelligence explosion “the Singularity,” which he designates as a time when “there will be no distinction between human and machine nor between physical and virtual reality” (9). For these thinkers, human intelligence will leave the sands and banks of human culture and ascend to loftier reaches beyond extinction.

Bostrom’s and Kurzweil’s rescue mission for human intelligence valorizes particular traits such as intelligence, reason, and autonomy, and simply sidesteps the entanglement of minds and bodies, of natural and human history. This narrow conception of human flourishing is customarily traced back to the Enlightenment. Ray Brassier’s *Nihil Unbound* radicalizes the relation between Enlightenment and extinction on philosophical grounds. Brassier holds that, if we are serious about wanting to pursue the truth, we should be happy to surrender the all too human inclination to find meaning in earthly things. Relinquishing such earthly attachments, for Brassier, releases new speculative possibilities. He invokes Lyotard’s though experiment, but dismisses Lyotard’s ultimate return to the body: solar catastrophe is an event that is by definition not witnessed, and in that way it shows that there are things that have meaning apart from what they mean for a human observer. Brassier follows philosopher Quentin Meillassoux in his opposition to “correlationism,” the idea, already touched on in Chapter 3, that there is “a necessary reciprocity between mind and nature” (Brassier 2007, 40). For Meillassoux, Brassier, and other so-called speculative realists, correlationism has been central to Western philosophy since Kant declared the “thing in itself” inaccessible for thought, and has curtailed the scope of philosophy by closing off “the great outdoors,” a reality where neither humans nor life are present (Meillassoux 2006, 26). For Brassier, human extinction can provide a paradoxical gateway to this uncharted cosmic terrain.

In contrast to Brassier, Meillassoux’s critique of correlationism does not hinge on a fantasy of extinction, but on a seemingly more benign thought experiment that takes us to a period *before* the dawn of human life. To prove that there is a reality that is unconstrained by its relation to humans, Meillassoux reminds us that science has yielded knowledge about things like the age of the universe, the date of the accretion of the earth, or the emergence of pre-human species. This means that it is possible to gauge the meaning of worlds that are strictly anterior to human consciousness. Meillassoux’s so-called “argument from ancestrality” focuses on what he calls “arche-fossils”: materials in our world that testify to the existence of such an ancestral world, such as radioactive isotopes allowing us to date fossils or the light emitted by a star informing us of its age (2006, 10). If we accept that such arche-fossils allow us to infer real knowledge about a world before us, and that such real knowledge constitutes the ultimate meaning about that world, then we must also accept that reality is not exhausted by the world as it appears to us. There is a world of wonder out there, just not for us (although Meillassoux is convinced that mathematical intuition can help us access it).

In Brassier (2007), Meillassoux's reenchantment of "the great outdoors" makes way for a dispassionate confrontation with futurelessness. Brassier prefers "the posteriority of extinction" to the wonders of ancestrality, because it more accurately conveys that all earthly concerns are always already pointless. In cosmic terms, human life has always already been over: "Extinction," he writes, "is not to be understood here as the termination of a biological species, but rather as that which *levels* the transcendence ascribed to the human" (224). The idea of extinction reminds us that the cosmos is "indifferent to our existence and oblivious to the 'values' and 'meanings' which we would drape over it in order to make it more hospitable" (xi). The rigor of Brassier's speculation, like the giddy excitement of Bostrom and Kurzweil, displays a studied refusal to get too excited about the waste of human lives. As Claire Colebrook (2018b) has noted, for these thinkers "the *loss of intellectual life as such* would be of a catastrophic order that far outweighs the tragedy of losing some or many humans" (162). This calculated indifference to suffering is a key aspects of these thinkers' intervention. We can call their refusal to recognize concerns over human survival as anything but banal and immature a "sado-dispassionate" stance (Plumwood 2002, 41–45): a posture that invokes emotive neutrality to blind itself to ethical concerns over forms of life that do not incarnate reason, autonomy, and intelligence. Such "calculated callousness" is a recurring feature of the imagining of "disanthropic" worlds, or worlds without humans (Garrard 2012, 45). It is not politically neutral: while some may argue that affective neutrality is the price we need to pay for philosophical rigor, philosophical arguments don't take place in a post-extinction wasteland, but in contexts that are marked by moral considerations, power differences, and emotive dispositions. Unleashing ideas of human superfluousness in volatile contexts can have consequences that are anything but speculative.

Misanthropy: Anti-Natalism, Exterminism, and Peripheral Life

Lofty declarations of thought's transcendence over earthly matters forget that human thought is currently actualizing itself as planetary destruction. Indifference to merely human catastrophes actively contributes to the destruction of lifeworlds. Greg Garrard has shown that visions of worlds without humans often collapse "into ordinary apocalypticism and ethical misanthropy" (2012, 49). We already encountered traces of misanthropy in post-apocalyptic fiction's aversion to great numbers, a hostility that precludes a necessary confrontation with the crucial but "morally obnoxious" theme of overpopulation (Clark 2016, 9). The most radical philosophical version of misanthropy is anti-natalism, which sees human birth as an intrinsically negative event, and believes there is a moral obligation to stop procreating. David Benatar (2006), the most notable philosophical voice defending this position, holds that "[b]ringing people into existence always inflicts serious harm on those people" (184). This allows for a conveniently unequivocal answer to "[t]he central question of population," which is the question of how many people there should be: "zero" (163–64). Because humans are a deeply flawed and destructive species, the amount of suffering in the cosmos would be dramatically diminished by getting rid of them. Benatar's calculus assumes a fundamental asymmetry between pleasure and pain, which assures that the latter always outweighs the former: "a life filled with good and containing only the most minute quantity of bad—a life of utter bliss adulterated only by the pain of a single pin-prick—is worse than no life at all" (48). And sure, "looming extinction would be bad for the final people," but they should be ready to take one for the team: after all, "we must take account not only of the final people's interests, but also of the harm that is avoided by not producing new generations" (198). Ideally, then, human life should go extinct as quickly as possible, if only to minimize the suffering that will inevitably accumulate until the last human dies.

Benatar's intervention is marked by a kind of deadpan philosophical rigor: a studiously sustained tone-deafness to human considerations. His misanthropic disdain for human life is absolute. Still, his insistence that his approach is philanthropic rather than misanthropic, as his argument is informed by the strictly utilitarian ambition to minimize human suffering, betrays a certain moral hesitation (223). Nor does Benatar propose that we actively exterminate the species: to that effect, he dismisses the notion of "lives worth living" (for him, not a single life meets that standard) and distinguishes between "lives worth continuing" and "lives worth starting" to argue that, while life is never worth starting, some lives that have already started are still worth continuing (22–28). So while there is apparently no moral imperative to go out and actively exterminate *all* human life ("dying-extinction" is better than "killing-extinction" [195]), Benatar allows for only two options: either more of the same or death. This myopic focus on (dis)continuation betrays a certain lack of imagination in Benatar's condemnation of human life. Like the transhumanism and speculative philosophy we explored in the previous section, and like a lot of post-apocalyptic fiction, Benatar only looks for meaning and value in modern life—it is just that he doesn't find it there. Like these other approaches, he fails to consider the possibility that other forms of life might be meaningful, as the likes of Haraway, Lerner, and McGuire do, and concludes that therefore human life should go extinct. The semi-serious Voluntary Human Extinction Movement (VHEMT), whose slogan "may we live long and die out" offers a more cheerful version of the diagnosis of human destructiveness, and it explicitly promotes human extinction as a way to, in Thom van Dooren's (2014) words again, "hold open space in the world for other living beings" (5). Other living beings, or other ways of living life, do not register on Benatar's radar.

For all its principled commitment to the all-encompassing worthlessness of human life, Benatar's anti-natalism implicitly valorizes a rational, autonomous, and intelligent form of existence over other forms. It is because the negative aspects of existence overshadow these values that human life deserves to go extinct; other dimensions of life cannot even claim to also be meaningful. Claire Colebrook (2018b) has remarked that Benatar's work also *performatively* affirms the values of autonomy and personhood: as a philosophical exercise in utilitarian calculation and rational decision-making, it ends up privileging the form of life that "is not so much calculated as *calculating*" (161). There is, for Colebrook, "a prima facie value placed on human capacity defined as rationality of a certain mode" (162). When this calculus is applied to a valuation of human life, it is unsurprising to see a distinction emerge between lives worth living and lives not worth living, and to find that the lives deemed worth living (or, in Benatar's bleaker worldview, most worth continuing life) are those with the power of calculation. In the contemporary world, this also means those with technology on their side.

Such hierarchies *between* human lives are mostly implicit in post-apocalyptic fiction, transhumanism, speculative realism, and anti-natalism, but they become all the more consequential in the disordered world of scarcity that is the Anthropocene. This is a world in which the very rich can afford a way out of climate calamities: a world in which residential real estate in cities like London is being bought up by global elites, not only for laundering money but also as a place of refuge in case of unrest in their home countries; in which the richest man alive, Jeff Bezos, is competing with other superstar CEO's like Elon Musk to colonize space, giving the lie to the slogan that there is no planet B; in which other Silicon Valley billionaires like Peter Thiel are buying up property in New Zealand to survive the collapse of civilization in their doomsday bunkers. In the scenarios that these power players anticipate, the lives of the rest of us become strictly superfluous.

In his reflections on the mutually enhancing effects of technological automation and climate crisis, Peter Frase (2016) has coined the term *exterminism* to name such doom scenarios. In a technologically advanced world, scarcity can be overcome for a tiny elite, and the result is a "communism for the few": a life of plenty that can, because of the finitude of our planetary resources, not be extended to the many (121). Thanks to technological developments, the lives of the happy few is rigorously independent from that of the rest of the population, who are needed neither as consumers (the riches of the few is consolidated, so normal cycles of production and consumption can be abandoned) nor producers (robots take care of that). As Frase argues, rampant global inequality has already put the necessary hierarchies in place, and the only thing missing today for this scheme to take effect is further technological development. If inequality was traditionally kept in check by the mutual dependence between capitalists and workers, in this scenario "an impoverished, economically superfluous rabble poses a great danger to the ruling class" without being of any potential use (123). Frase fears that this will lead Malthusian fears to flip over into a program of extermination. The increasing spatial differentiation between the gated communities of the rich and the housing of the rest of the population, the proliferation of zones where refugees and prisoners effectively live in concentration camps, and the increasing militarization of police forces make such derailments practically possible. This would be the outcome of a toxic cocktail of a misanthropic devaluation of human life, extreme inequality, and technological development.

In this book, literature has done many things: it has described Anthropocene realities, imagined possible futures, remixed different kinds of disciplinary knowledge, and rescaled overwhelming realities. It has also provided imaginative devices, such as the figure of a posthumous reader, animal symbionts, or seas magically scribbling Wordsworth poems, that enrich debates over the human impact on the environment. Frase also sees his own project in his book *Four Futures*, in which he envision four possible futures for a climate changed planet, as deeply informed by literature: he sees it as a form of "social science fiction" that combines the techniques of speculative fiction with the insights of the social sciences in the ecological future of the planet (24). Frase's mix of literary and scientific affordances to address the ecological crisis exemplifies the mandate of the environmental humanities: a project of interdisciplinary description that keeps different differences—be they cultural, social, ontological, or otherwise—in focus. Frase finds the traditional distinction between "soft" and "hard" science fiction less relevant than that between "stories that take their world-building seriously, and those that don't" (25). I have argued that it is through world-building, and by exploring the limitations and dangers of world-building, that literature can upgrade the project of the environmental humanities.

In the introduction, I singled out William Gibson's 2014 novel *The Peripheral* as a work whose technological, economic, and social world-building intervenes in Anthropocene debates. As it directly engages with the horrors and fantasies of human extinction, it will here serve as my last literary example. The novel imagines a bifurcated future: a near future situated in an impoverished USA, and a twenty-second-century future inhabited by a small elite living technologically enhanced lives. While information can travel back in time, physical matter cannot. The "hard" science fiction explanation for this phenomenon (an obscure process of "quantum tunneling" through a Chinese computer server) is less relevant than the way this set-up shapes the power differences between the two worlds: the hi-tech future can freely intervene in the near future, and the games the people in the near future are paid to play provide actual labor in the later future. This is enabled by superior data-processing technology: "Information from there affects things here ... Their stuff's all seventy years faster than ours" (Gibson 2015, 192). This exploitation of the earlier future operates without fear of retribution or upheaval: once the future world connects with and interferes in the past, that past stops being their past and becomes an alternative timeline, "a stub" (38). This is why Gibson has referred to the novel's near future as a "third worlded version of contemporary America" (Newitz 2014): the same impudence with which colonial powers extracted, and continue to extract, labor and natural resources from the Global South is now unleashed against the American population through data processing.

Gibson famously wrote the first classic cyberpunk novel with *Neuromancer* in 1984. And if the notion of the Anthropocene was not yet coined then, the first sentence of that novel already betrays an interest in the thorough interpenetration of the human, the natural, and the technological: "The sky above the port was the color of television, tuned to a dead channel" (Gibson 2004, 3). In *The Periperal*, the Anthropocene is imagined as "the Jackpot," a multicausal, slow, drawn-out collapse of civilization initiated by climate change, which also results in political destabilization, mass extinction, and the end of democracy. The Jackpot separates the novel's two futures, and it kills about 80 percent of the world population over about 40 years (320). But unlike what the term may suggest, the outcome of the Jackpot is not entirely arbitrary, as the rich manage to turn "constant crisis" into "constant opportunity." The 80 percent are merely collateral damage for a shiny new world "lit increasingly by the new" (321). For the "[o]ligarchs, corporations, neo-monarchists" surviving the Jackpot, the Anthropocene is "[a] progress accompanied by constant violence" (321, 322). It inaugurates what Frase calls "a communism for the few," who have not bothered to save excess populations, as they can count on infinite labor reserves in alternative "stubs" that will never dry up. Only because, in the world of the novel, past lives are cheaper than technology and do not constitute a physical threat are they allowed to remain alive. The novel's title not only refers to one of the novel's hi-tech gimmicks, a kind of drone body that people can inhabit from a distant location, but it also indicates the status of most lives in this new dispensation: the majority of lives are marginal, of secondary or only superficial importance.

The Peripheral traces the overlaps between transhumanist fantasy and misanthropic indifference in a way that enriches debates over Anthropocene extinction in different ways. It paints a fine-grained picture (Gibson calls it "really high resolution SF" [Kross 2014]) of two faces of the near future: the derelict trailer-park reality of impoverished, abandoned, and uneducated excess populations, and the contrasting glamor and splendor of the post-Jackpot "klept." The asymmetrical relation between the two futures not only drives the novel's thriller plot, it also allows the novel to narrate a relation that clearly reflects the rampant inequality in our present. The decision to position the Jackpot in between these two moments places it at the heart of contemporary power relations, and allows the novel's narrative to circumscribe the terms of what Gibson calls the "unspeakable present" (Kross 2014).

This asymmetrical relation between the two futures also resonates with the affinities between the Anthropocene and writing that I discussed in Chapter 3. A number of characters in the far future feel deeply nostalgic about the "[g]loriously pre-posthuman" times before the Jackpot (75). Paradoxically, they learn the hard lesson that inscriptions cannot be erased, and past deeds cannot be undone, through their power to interfere with the past. In the world of the novel, the past can be altered, but the cruel point is that, if the past is successfully interfered with, it stops being part of the future "continuum," and is relegated to the status of a mere "stub" that is strictly irrelevant to the fate of the world. Erasure, in *The Peripheral*, is only possible at the price of irrelevance. This might be the novel's (and this book's) final word on the Anthropocene: if human life wants to continue to matter, it needs to resist fantasies of erasure and innocence and face up to its painful and violent legacies. To return to the sandy imagery with which I began this chapter, this amounts to the choice between assuming Ozymandias's "shattered visage" or escaping into the evasive comforts of Foucault's "face drawn in the sands." Offering such relief, this book has argued, is not what literature is for.

Glossary

Affect A term that refers to sensations, experiences, and dispositions that cannot be fully captured by an individual's consciousness and cognition, but that emerge in the encounter between bodies, minds, and the outer world. In affect theory, affects differ from emotions, which are typically considered to be *owned* by human individuals: people *have* emotions, while affects circulate *between* people, texts, and things. Tracing how literary texts register affect can point us to emergent or submerged dimensions of life that have not yet been consolidated in stable stories and concepts.

Affordance In design theory, an object's affordances name the different uses and actions that a design or material make possible. An affordance does not constrain or compel, it merely enables and facilitates. Caroline Levine has adopted the term for literary studies, where it provides a flexible way for describing how literary forms act on their readers.

Agency Traditionally refers to the capacity of individuals to make free choices, act independently, and have a real-world impact. Agency was often contrasted to structure, which is assumed to constrain individuals' agency, but the term has increasingly come to refer to the power of human and nonhuman actors alike to affect the world they compose. In this sense, agency no longer implies autonomy and free will, but is always distributed between different actors.

Apocalypse Refers to the end of the world, and conceives of that moment as a revelation of an ultimate truth. In a religious framework, apocalypse signals the demise of earthly existence and the advent of an otherworldly reality. In secular contexts, apocalyptic ideas often channel anxieties about the fragility and impermanence of the institutions and ideals that sustain our lives. Apocalyptic literature has a paradoxical temporal structure, as telling about the apocalypse implies that the cataclysmic event has not been fatal—has, in a sense, not been truly apocalyptical.

Assemblage Names the process of organizing, arranging, and fitting together that results in what Jane Bennett calls "ad hoc groupings of diverse elements." Typically, the assemblage as a whole has a power and an agency that differ from that of the different entities, bodies, qualities, and expressions it brings together, and that are not controlled by any one of the composite parts.

Cli-fi An increasingly ubiquitous literary genre that deals with climate change and global warming. Cli-fi, like sci-fi (science fiction), is customarily set in the future, but as the effects of climate change are becoming ever more apparent, cli-fi often no longer needs to imagine a non-contemporary world.

Cosmopolitanism The belief in and cultivation of a sense of solidarity and community that extends beyond national borders and may encompass all inhabitants of the world. Cosmopolitanism often goes hand in hand with the conviction that literature and art can foster a sense of intercultural connection. In the context of the global ecological crisis, promoting a sense of environmental world citizenship or "eco-cosmopolitanism," as Ursula Heise calls it, may be indispensable for developing a concerted response to the threat of a warming planet.

Critter Officially a general (if often disparaging) term to refer to animals, the notion is used by Donna Haraway to indicate the whole range of life forms with which human life, according to her, has to "make kin" in the Anthropocene. Referring promiscuously to plants, animals, microbes, humans, nonhumans, and even machines, the term emphasizes the multiplicity of the affiliations through which human life is constituted.

Cybernetics A research project that aims to provide an encompassing account of how humans, animals, and machines interact with one another. Cybernetics studies the interactions between different entities as a closed system in which information circulates, with the aim of controlling that circulation. It flourished in a Cold War context where computers were mobilized for the wholesale militarization and control of the planet.

Deconstruction A way of doing philosophy and of reading (mostly literary) texts that aims to destabilize and complicate the relation between text and meaning. Deconstruction emphasizes that texts often generate an uncontrollable excess of meanings that analysis can never conclusively determine. Deconstruction was very popular in literature departments in the 1970s and 1980s, before it made way for approaches that recovered a more robust relation between language and the world.

Defamiliarization Viktor Shklovsky's term for literature's capacity to provide surprising new perspectives on the world. A translation of the Russian "*ostranenie*," defamiliarization (or estrangement or even enstrangement) accentuates the uses of literature in interrupting customary ways of viewing and inhabiting the world.

Disanthropy Different from misanthropy, which refers to a dislike or hatred of other humans, this notion was deployed by Greg Garrard to name a desire for a world without humans. Disanthropy constitutes a formal challenge, as most literary and artistic forms and genres imply a human voice, character, or perspective, which complicates the representation of (often far future) worlds without us.

Discorrelation A term used by Shane Denson to refer to technological processes that liberate cultural products, such as cinema, from the limits of human perception and cognition. An example is the continuous bombardment of hyperinformatic images in recent action cinema, which outstrips human perceptual faculties. The term resonates with Quentin Meillassoux's notion of "correlationism," which conveys that philosophical thought since Kant has restricted its remit to the relation between mind and matter. Processes of discorrelation can arguably step outside this circle and offer a glimpse of "the great outdoors."

Dystopia A genre of speculative fiction depicting an undesirable and frightening future. Twentieth-century dystopias typically represent a totalitarian world where technology is used to curtail individual autonomy. In the twenty-first century, dystopias more often evoke a diminished, impoverished, and brutalized world on a climate changed planet. Often, dystopias exaggerate tendencies that are already latent in contemporary society to warn their audiences about emergent threats to our ways of life.

Earthbound Bruno Latour's name for an alliance of political actors who are willing to confront climate change. The Earthbound realize that nature is not simply a resource that humans can use up and abandon (bound *to* the Earth) and that political action is essentially concerned with imagining a more sustainable relation to the planet (bound *for* a different relation to the Earth, or what Latour calls the "New Climate Regime"). The Earthbound find themselves at war with defenders of the status quo whom Latour calls Humans.

Ecomodernism The conviction that a civilization that has destroyed ethnic minorities on an industrial scale and plundered the planet to enrich a small elite can be trusted to solve the ecological crisis through technological progress and the miraculous operations of the free market.

Entanglement Term that refers to the deep interconnections between human and nonhuman actors, as well as between material and meaning-making processes, to the point that these can no longer be separated and distinguished. For Karen Barad, whose use of the term is inspired by quantum physics, entanglement means that it no longer makes sense to even speak of independent, self-contained entities: reality is entangled all the way down. The term has become unavoidable in contemporary materialist thought.

E-waste Names discarded electric and electronic devices. This waste often remains out of view when contemporary information technology is celebrated as clean, fast, and immaterial. Electronic scrap components are often extremely toxic, and they retain their toxicity for a very long time. And e-waste is dirty in a second sense: it exposes the workers in developing countries responsible for dismantling and disposing of electronic waste to extreme health and safety risks.

Future anterior A grammatical tense that expresses a future event that precedes another future event ("the human species will have left a trace on the surface of the Earth"). Mark Currie has emphasized that the future anterior also lets us apprehend our present as a future memory, as something that will have been. The future anterior is central to the Anthropocene: it conveys the idea that the human impact on the planet will be readable in the future, and it imbues our sense of the future with a memorial, past-oriented, and even mournful dimension.

Gaia Names a thesis, developed by James Lovelock and Lynn Margulis, that the Earth is a complex self-regulating system in which living organisms work to maintain the environmental conditions for life. The idea combines ecological and cybernetic insights, and it has had a huge influence on environmental thought. In his revision of the notion, Bruno Latour emphasizes that Gaia is not a closed totality, and is in reality an unruly anti-system.

Hermeneutics The theory and methodology of interpreting texts and, more generally, finding meaning in particular texts, actions, psychological phenomena, and social situations. Traditionally applied in the study of sacred texts, hermeneutics has in modern times been developed in different philosophical and disciplinary contexts. In an Anthropocene context, hermeneutics points to the challenge of reading signals and indications of the human impact on the planet.

Humanism The philosophical position that emphasizes the value of human life in general, as opposed to the nonhuman world, and that underlines individual ideals such as autonomy, rationality, and independence. The emphasis on human flourishing makes humanism an essentially non-religious, secular development, but the normative insistence on particular values and the disregard for nonhuman entities has led to criticism from posthumanist and materialist perspectives.

Hyperobject Timothy Morton's term for entities such as plutonium, global warming, and thermoplastics that are distributed in time and space to the point that we cannot apprehend them as particular objects anymore. Some contemporary philosophies see the overwhelming and disorienting character of hyperobjects as key characteristics of the object world as such.

Inscription In the most straightforward sense, the material process and the product of writing something on a particular surface. The term foregrounds the hope for durability and the physical displacement inherent in such actions, and this leads to a more technical sense, in which inscription becomes a model for human interactions with the planet: impossible to fully erase, at risk of turning violent, and displaying a tension between meaning-making and physical force.

Lifeworld philosophical notion that refers to the self-evident reality that individuals and groups inhabit, and that imperceptibly forms the horizon within which they create meaning. In an environmental context, lifeworld (*Lebenswelt*) often merges with the notion of *Umwelt* to name the environment, the habitat in which human and nonhuman entities live.

Litany Originally a form of prayer in which the clergy leads and the congregation responds in a repetitive pattern. The term is used more generally to refer to a repetitive and lengthy enumeration, and then its meaning comes close to that of a catalogue. The form of the litany or the catalogue is a key component of some kinds of critical engagement with the Anthropocene world, which evoke the multiplicity and excess of that world through long enumerations.

Materiality In the most straightforward way, this notion refers to the physical properties of things, often to oppose tendencies that privilege the ideational or linguistic dimensions of reality. In recent materialist thought, however, the distinction between these different realms is often superseded by the insistence that meanings and signs are also thoroughly material. In deconstructionist thought, the notion of materiality even refers to those aspects of reality that resist appearing as part of the phenomenal world.

Mesh Timothy Morton's term for the infinitely complex interrelations between humans and a whole series of nonhuman entities, ranging from tiny bacteria to vast realities such as climate change. The mesh is a radically open form without center or edge, and it foregrounds the interdependence of different living and nonliving things.

Monism The philosophical notion that reality consists of a single reality or substance, and that at the deepest level of reality, there is no distinction between different kinds of entities, not even between tangible objects and mental states, between things and consciousness. Monism is a key characteristic of materialist philosophy, and it is generally opposed to dualism, which, most famously in the thought of René Descartes, holds that mind and body are distinct and separable.

Multidirectional memory A term coined by Michael Rothberg to capture the fact that the shared cultural memories of particular groups do not exist in isolation from one another in a kind of silo, but are constituted, updated, and revised through intense interactions between different memory traditions. Rosanne Kennedy has coined the notion of "multidirectional eco-memory" to highlight that this interactive dynamics can also engage with environmental contexts, and not only with the manmade violence in relation to which Rothberg coined the notion.

Naturalization Refers to a rhetorical process in which nature imagery is used to present social and cultural structures as if they were part of nature. Generally, naturalization tends to present contingent and constructed realities as if they were self-evident and changeless, and thus discourages contestation of those realities.

Natureculture Donna Haraway's notion expressing that biological (both human and nonhuman) and technological forces constitute one another through their constant complex interactions. This fundamental relatedness makes it impossible to separate history and biology, culture and nature.

Ontology The branch of philosophy that studies the basic nature or essence of everything that exists. The term is also often used to describe a particular account of the kind of properties and relations that pertain in reality, or in a sector of reality (as when we speak of, for instance, the social ontology of capitalism). Since the beginning of the century, critical thought has increasingly focused on the description of reality at a basic level, rather than on the limits of language. In several fields, this shift has been called an "ontological turn."

Petromelancholia A term coined by Stephanie LeMenager to name an affective and imaginative inability to let go of the pleasures of lifestyles that are only possible through the easy availability of cheap energy. Now that we know our dependence on these resources is not sustainable, petromelancholia often obstructs the design of different forms of life. The notion is related to Lauren Berlant's account of "cruel optimism," which names our desperate desire for something that is actually an obstacle to our flourishing.

Reproductive futurism The belief that what is ultimately at stake in politics is creating a better future for our children. According to Lee Edelman, this conviction puts forward heterosexuality as the only acceptable way of life and leaves no room for queer sexualities. A queer politics then consists in a refusal of the future and of the deified figure of the child.

Scriptorium A room in a medieval monastery dedicated to writing, and especially for copying and illuminating manuscripts. Such a space committed to the practice of writing offers an apt image for the way the Anthropocene world consists of the accumulated traces and inscriptions of human and nonhuman actions, which are compulsively read, pondered, and interpreted, often with the help of technology.

Semiosphere Just as the biosphere names the zone in which life operates and sustains itself, this notion, coined by Yuri Lotman, names the environment in which signs are produced and read and in which meaning is generated. The notion emphasizes the power of human and nonhuman actors to co-create the environments in which they live. In some new materialist thinking, the semiosphere and the biosphere become virtually synonymous, as all living entities are endowed with the capacity to create meaning, and as all activities can be seen as forms of signification, inscription, and meaning-making.

Speculative In philosophy, a mode of thinking that moves beyond empirical evidence and tries to find access to a transcendent reality—traditionally, to the divine or the absolute, but in contemporary speculative realism, to the deep reality of things, which is not captured by human perception or customary modes of cognition. In literature, speculative fiction tells stories that are not restricting themselves to the depiction of a realistic world, but that import futuristic, supernatural, and virtual elements. Science fiction, fantasy, and horror are prominent forms of speculative fiction.

Sublime In contrast to the beautiful, which is conceived as orderly and harmonious and provides a mild pleasure, this aesthetic category names an experience in which the human is overwhelmed by the power, the size, or the multiplicity of sounds and visions. This experience inspires awe and reverence, and ultimately aggrandizes rather than diminishes the human individual, who proves him- or herself capable of transcending merely rational thinking. It is this recuperative dimension of the sublime that has occasioned most criticism.

Symbiont An organism that participates in a process of symbiosis, that is, in a cooperative or very intimate relationship between at least two different organisms. The term is used by multispecies thinker Donna Haraway to stress the mutually beneficial nature of the relations and collaborations between humans and different nonhuman species.

Trans-corporeality Is the idea that all creatures are affected by the dynamics of the material world. Stacy Alaimo, who coined the term, situates it in a tradition of feminist thinking that opposes the idea that the human is disembodied, discrete, and detached, and that instead underscores that human life is always embodied, exposed, and permeable to outside forces.

Transhumanism The belief that the bodily and cognitive limitations of human life can be overcome through the thorough integration of human life with technology. This techno-optimist perspective looks forward to a future when the merger of artificial intelligence, prostheses, and human lives will generate life forms that are not constrained by the biological limitations that currently still define human existence.

Weird Literary genre that evokes dread and horror by confronting cosmic forces that reduce human life to insignificance. The sheer strangeness and tolerance for ludicrousness that marks this literary tradition has made it a productive mode for contemporary environmental writing that seeks to confront the bewildering changes that the Anthropocene brings about. Often, "new weird" fiction has a more welcoming and accommodating attitude to a mysterious and monstrous reality.

Wilderness A natural domain that is not contaminated by culture. As William Cronon has shown, the notion of a wilderness emerges from within a culture that finds itself alienated from the natural world, and not with people who maintain a more nourishing relation to the natural world. Historically, the notion of the wilderness has fueled a desire to conquer and subdue nature; more recently, cultivating a sense of unspoiled nature often prevents the designing of livable interactions between human and nonhuman life.

References

Ackerman, Diane. *The Human Age: The World Shaped by Us*. New York: Norton, 2014.

Adamson, Joni, and Scott Slovic. "The Shoulders We Stand On: An Introduction to Ethnicity and Ecocriticism." *Melus* 34. 2 (2009): 5–24.

Alaimo, Stacy. *Bodily Natures: Science, Environment, and the Material Self*. Bloomington: Indiana University Press, 2010.

Alaimo, Stacy. "Oceanic Origins, Plastic Activism, and New Materialism at Sea." *Material Ecocriticsm*. Edited by Serenella Iovino and Serpil Oppermann. Bloomington: Indiana University Press, 2014. 186–203.

Alaimo, Stacy. *Exposed: Environmental Politics and Pleasures in Posthuman Times*. Minneapolis: University of Minnesota Press, 2016.

Albrecht, Glenn. "'Solastalgia': A New Concept in Health and Identity." *PAN: Philosophy Activism Nature* 41. 3 (2005): 41–55.

Appel, Hannah. "Infrastructural Time." *The Promise of Infrastructure*. Edited by Hannah Appel, Nikhil Anand, and Akhil Gupta. Durham, NC: Duke University Press, 2018. 41–61.

Appel, Hannah, Nikhil Anand, and Akhil Gupta. "Introduction: Temporality, Politics, and the Promise of Infrastructure." *The Promise of Infrastructure*. Edited by Hannah Appel, Nikhil Anand, and Akhil Gupta. Durham, NC: Duke University Press, 2018. 1–38.

Arendt, Hannah. *The Human Condition*. 2nd edition. Chicago, IL: University of Chicago Press, 1998.

Armstrong, Nancy. *How Novels Think: The Limits of Individualism from 1719–1900*. New York: Columbia University Press, 2005.

Arrighi, Giovanni. *The Long Twentieth Century: Money, Power, and the Origins of Our Times*. London: Verso, 1995.

Asafu-Asjaye, John, et al. "An Ecomodernist Manifesto." Ecomodernism.org, Apr. 2015. Web.

Assis, Claudia. "Elon Musk on SpaceX Launch: 'Moon was in the way'." Market-Watch.com, 14 Apr. 2015. Web.

Augé, Marc. *Non-Places: An Introduction to Supermodernity*. London: Verso, 1995.

Baccolini, Raffaella, and Tom Moylan. "Introduction: Dystopia and Histories." *Dark Horizons: Science Fiction and the Dystopian Imagination*. Edited by Raffaella Baccolini and Tom Moylan. New York: Routledge, 2003. 1–12.

Barad, Karen. *Meeting the Universe Halfway: Quantum Physics and the Entanglement of Matter and Meaning*. Durham, NC: Duke University Press, 2007.

Baucom, Ian. "The Human Shore: Postcolonial Studies in an Age of Natural Science." *History of the Present* 2. 1 (2012): 1–23.

Beck, Ulrich. "Emancipatory Catastrophism: What Does It Mean to Climate Change and Risk Society?" *Current Sociology* 63. 1 (2015): 75–88.

Benatar, David. *Better Never to Have Been: The Harm of Coming into Existence*. Oxford: Oxford University Press, 2006.

Benjamin, Walter. "On the Concept of History." *Selected Writings Vol. 4: 1938–40*. Cambridge, MA: Harvard University Press, 2006. 389–400.

Bennett, Jane. *Vibrant Matter: A Political Ecology of Things*. Durham, NC: Duke University Press, 2010.

Berger, James. *After the End: Representations of Post-Apocalypse*. Minneapolis: University of Minnesota Press, 1999.

Berlant, Lauren. *The Female Complaint: The Unfinished Business of Sentimentality in American Culture*. Durham, NC: Duke University Press, 2008a.

Berlant, Lauren. "Intuitionists: History and the Affective Event." *American Literary History* 20. 4 (2008b): 845–860.

Best, Stephen, and Sharon Marcus. "Surface Reading: An Introduction." *Representations* 108 (2009): 1–21.

Blum, Hester. "Melville and Oceanic Studies." *The New Cambridge Companion to Herman Melville*. Edited by Robert S. Levine. Cambridge: Cambridge University Press, 2014. 22–36.

Boes, Tobias, and Kate Marshall. "Writing the Anthropocene." *Minnesota Review* 83 (2014): 60–72.

Bond, Lucy, Ben De Bruyn, and Jessica Rapson. "Planetary Memory in Contemporary American Fiction." *Textual Practice* 31. 5 (2017): 853–866.

Bonneuil, Christophe, and Jean-Baptiste Fressoz. *The Shock of the Anthropocene*. London: Verso, 2016

Bostrom, Nick. "Introduction—The Transhumanist FAQ: A General Introduction." *Transhumanism and the Body: The World Religions Speak*. Edited by Calvin Mercer and Derek Maher. New York: Palgrave Macmillan, 2014a. 1–17.

Bostrom, Nick. *Superintelligence: Paths, Dangers, Strategies*. Oxford: Oxford University Press, 2014b.

Boyer, Dominic. "Dominic Boyer on the Anthropology of Infrastructure." Blog. castac.org, 3 Mar. 2014. Web.

Boyer, Dominic. *Energopolitics: Wind and Power in the Anthropocene*. Durham, NC: Duke University Press, 2019.

Braidotti, Rosi. *The Posthuman*. Oxford: Polity, 2013.

Brassier, Ray. *Nihil Unbound: Enlightenment and Extinction*. Basingstoke: Palgrave Macmillan, 2007.

Buell, Lawrence. *The Environmental Imagination: Thoreau, Nature Writing, and the Formation of American Culture*. Cambridge, MA: Harvard University Press, 1995.

Buell, Lawrence. *The Future of Environmental Criticism: Environmental Crisis and Literary Imagination*. Malden, MA: Blackwell, 2005.

Buell, Lawrence. "Ecoglobalist Affects: The Emergence of U.S. Environmental Imagination on a Planetary Scale." *Shades of the Planet: American Literature as World Literature*. Edited by Wai Chee Dimock and Lawrence Buell. Princeton, NJ: Princeton University Press, 2007. 227–248.

Buell, Lawrence. "Ecocriticism: Some Emerging Trends." *Qui Parle* 19. 2 (2011): 87–115.

Burnside, John. *Glister*. London: Vintage, 2009.

Canavan, Gerry. "Memento Mori: Richard McGuire's *Here* and Art in the Anthropocene." Deletionscifi.org, 5 Nov. 2017. Web.

Carson, Rachel. *Silent Spring: Fortieth Anniversary Edition*. New York: Houghton Mifflin, 2002.

Chakrabarty, Dipesh. "The Climate of History: Four Theses." *Critical Inquiry* 35. 2 (2009): 197–222.

Chakrabarty, Dipesh. "Postcolonial Studies and the Challenge of Climate Change." *New Literary History* 43. 2 (2012): 1–18.

Cheah, Pheng. *What Is a World? On Postcolonial Literature as World Literature*. Durham, NC: Duke University Press, 2016.

Clark, Timothy. "Some Climate Change Ironies: Deconstruction, Environmental Politics, and the Closure of Ecocriticism." *Oxford Literary Review* 32. 1 (2010): 131–149.

Clark, Timothy. *Ecocriticism on the Edge: The Anthropocene as a Threshold Concept*. London: Bloomsbury, 2015.

Clark, Timothy. "'But the real problem is....': The Chameleonic Insidiousness of 'Overpopulation' in the Environmental Humanities." *Oxford Literary Review* 38. 1 (2016): 7–26.

Clarke, Michael Tavel, and David Wittenberg. "Introduction." *Scale in Literature and Culture*. Edited by Michael Tavel Clarke and David Wittenberg. Cham: Palgrave Macmillan, 2017. 1–32.

Cole, Teju. *Open City*. London: Faber & Faber, 2011.

Colebrook, Claire. "Introduction: Extinction. Framing the End of the Species." *Extinction*. Edited by Claire Colebrook. *Living Books about Life*, 2012. Web.

Colebrook, Claire. *The Death of the PostHuman: Essays on Extinction, Volume One*. Ann Arbor, MI: Open Humanities Press, 2014.

Colebrook, Claire. "We Have Always Been Post-Anthropocene: The Anthropocene Counterfactual." *Anthropocene Feminism*. Edited by Richard Grusin. Minneapolis: University of Minnesota Press, 2017. 1–20.

Colebrook, Claire. "Extinction." *Posthuman Glossary*. Edited by Rosi Braidotti and Maria Hlavajova. London: Bloomsbury, 2018a. 150–153.

Colebrook, Claire. "Lives Worth Living: Extinction, Persons, Disability." *After Extinction*. Edited by Richard Grusin. Minneapolis: University of Minnesota Press, 2018b. 151–171.

Colebrook, Claire. "The Future in the Anthropocene: Extinction and the Imagination." *Climate and Literature*. Edited by Adeline Johns-Putra. Cambridge: Cambridge University Press, 2019. 263–280.

Craps, Stef, Rick Crownshaw, Jennifer Wenzel, Rosanne Kennedy, Claire Colebrook, and Vin Nardizzi. "Memory Studies and the Anthropocene: A Roundtable." *Memory Studies* 11. 4 (2018): 498–515.

Cronon, William. "The Trouble with Wilderness, or, Getting Back to the Wrong Nature." *Environmental History* 1. 1 (1996): 7–28.

Crownshaw, Richard. "Climate Change Fiction and the Future of Memory: Speculating on Nathaniel Rich's *Odds against Tomorrow*." *Resilience* 4. 2–3 (2017a): 127–146.

Crownshaw, Richard. "Cultural Memory Studies in the Epoch of the Anthropocene." *Memory Unbound: Tracing the Dynamics of Memory Studies*. Edited by Lucy Bond, Stef Craps, and Pieter Vermeulen. New York: Berghahn, 2017b. 242–257.

Crutzen, Paul. "Geology of Mankind." *Nature* 415 (2002): 23.

Crutzen, Paul, and Christian Schwägerl. "Living in the Anthropocene: Toward a New Global Ethos." Yale Environment 360, 24 Jan. 2011. Web.

Cubitt, Sean. *Finite Media: Environmental Implications of Digital Technologies*. Durham, NC: Duke University Press, 2017.

Cunsolo, Ashlee, and Neville Ellis. "Ecological Grief as a Mental Health Response to Climate Change-Related Loss." *Nature Climate Change* 8 (2018): 275–281.

Danowski, Déborah, and Eduardo Viveiros De Castro. *The Ends of the World*. Malden, MA: Polity, 2017.

Darwin, Charles. *On the Origin of Species: A Facsimile of the First Edition*. Cambridge, MA: Harvard University Press, 2001.

Davis, Heather. "Life & Death in the Anthropocene: A Short History of Plastic." *Art in the Anthropocene: Encounters among Aesthetics, Politics, Environments, and Epistemologies*. Edited by Heather Davis and Etienne Turpin. London: Open Humanities Press, 2015. 347–358.

Davis, Kimberly Chabot. "The Follies of Racial Tribalism: Mat Johnson and Anti-Utopian Satire." *Contemporary Literature* 58. 1 (2017): 18–52.

De Boever, Arne. *Plastic Sovereignties: Agamben and the Politics of Aesthetics*. Edinburgh: Edinburgh University Press, 2016.

De Bruyn, Ben. "Realism 4°: Objects, Weather, and Infrastructure in Ben Lerner's *10:04*." *Textual Practice* 31. 5 (2017): 951–971.

De Bruyn, Ben. *The Novel and the Multispecies Soundscape*. Cham: Palgrave Macmillan, 2020.

DeLanda, Manuel. *War in the Age of Intelligent Machines*. New York: Zone Books, 1991.

DeLillo, Don. *Point Omega*. London: Picador, 2011.

DeLoughrey, Elizabeth. "Submarine Futures of the Anthropocene." *Comparative Literature* 69. 1 (2017): 32–44.

DeLoughrey, Elizabeth. *Allegories of the Anthropocene*. Durham, NC: Duke University Press, 2019.

de Man, Paul. *The Resistance to Theory*. Minneapolis: University of Minnesota Press, 1986.
Demos, Thomas J. *Against the Anthropocene: Visual Culture and Environment Today*. Berlin: Sternberg Press, 2017.
Denson, Shane. "Post-Cinema after Extinction." *Media Fields Journal* 13 (2018): 1–13.
Denson, Shane, and Julia Leyda. "Perspectives on Post-Cinema: An Introduction." *Post-Cinema: Theorizing 21st-Century Film*. Edited by Shane Denson and Julia Leyda. Sussex: REFRAME Books, 2016. 1–20.
Derrida, Jacques. *Of Grammatology*. Baltimore, MD: Johns Hopkins University Press, 1974.
Derrida, Jacques. "Archive Fever: A Freudian Impression." *Diacritics* 25. 2 (1995): 9–63.
Derrida, Jacques. *Rogues: Two Essays on Reason*. Stanford, CA: Stanford University Press, 2005.
Dillon, Sarah. "Imagining Apocalypse: Maggie Gee's *The Flood*." *Contemporary Literature* 48. 3 (2007): 374–397.
Dorrian, Mark. "Adventure on the Vertical: *Powers of Ten* and the Mastery of Space by Vision." *Cabinet* 44 (2011): 17–22.
Edelman, Lee. *No Future: Queer Theory and the Death Drive*. Durham, NC: Duke University Press, 2004.
Edwards, Paul. *A Vast Machine: Computer Models, Climate Data, and the Politics of Global Warming*. Cambridge, MA: MIT Press, 2010.
Ehrlich, Paul. *The Population Bomb*. Cutchogue: Buccaneer Books, 1971.
Emmett, Robert, and David Nye. *The Environmental Humanities: A Critical Introduction*. Cambridge, MA: MIT Press, 2017.
Fay, Jennifer. *Inhospitable World: Cinema in the Time of the Anthropocene*. Oxford: Oxford University Press, 2018.
Felski, Rita. *The Limits of Critique*. Chicago, IL: University of Chicago Press, 2015.
Foucault, Michel. *The Order of Things: An Archeology of the Human Sciences*. New York: Routledge, 2002.
Frase, Peter. *Four Futures: Visions of the World after Capitalism*. London: Verso, 2016.
Fuller, Buckminster. *Your Private Sky: Discourse*. Baden: Lars Müller, 2001.
Garrard, Greg. *Ecocriticism*. London: Routledge, 2004.
Garrard, Greg. "Worlds without Us: Some Types of Disanthropy." *Substance* 41. 1 (2012): 40–60.
Ghosh, Amitav. *The Great Derangement: Climate Change and the Unthinkable*. Chicago, IL: University of Chicago Press, 2016.
Gibson, William. *Neuromancer*. New York: Penguin, 2004.
Gibson, William. *The Peripheral*. London: Penguin, 2015.
Gilroy, Paul. *The Black Atlantic: Modernity and Double-Consciousness*. Cambridge, MA: Harvard University Press, 1995.
Gleeson-White, Jane. "Country and Climate Change in Alexis Wright's *The Swan Book*." *AJE: Australasian Journal of Ecocriticism and Cultural Ecology* 6 (2017): 29–38.

Gourevitch, Philip. "'There is another world in our world': A Conversation with J. M. Ledgard." *New Yorker*, 30 Apr. 2013. Web.

Gratton, Peter. *Speculative Realism: Problems and Prospects*. London: Bloomsbury, 2014.

Grove, Jairus Victor. *Savage Ecology: War and Geopolitics at the End of the World*. Durham, NC: Duke University Press, 2019.

Grusin, Richard, editor. *The Nonhuman Turn*. Minneapolis: University of Minnesota Press, 2015.

Haldane, J.B.S. "On Being the Right Size." *The World of Mathematics Vol. 2*. Edited by James Newman. Mineola: Dover, 1956. 952–958.

Haraway, Donna. "A Cyborg Manifesto: Science, Technology, and Socialist-Feminism in the Late Twentieth Century." *Simians, Cyborgs, and Women: The Reinvention of Nature*. New York: Routledge, 1991. 149–181.

Haraway, Donna. *The Companion Species Manifesto: Dogs, People, and Significant Otherness*. Chicago, IL: University of Chicago Press, 2003.

Haraway, Donna. *Staying with the Trouble: Making Kin in the Chthulucene*. Durham, NC: Duke University Press, 2016.

Haraway, Donna, et al. "Anthropologists Are Talking—About the Anthropocene." *Ethnos* 81. 3 (2016): 535–564.

Harbach, Chad. "The End." *N+ 1* 6 (2008). Web.

Harman, Graham. *Guerrilla Metaphysics: Phenomenology and the Carpentry of Things*. Peru, IL: Open Court, 2005.

Harman, Graham. *The Quadruple Object*. Winchester: Zero Books, 2011.

Harman, Graham. "The Well-Wrought Broken Hammer: Object-Oriented Literary Criticism." *New Literary History* 43. 2 (2012): 183–203.

Harman, Graham. *Immaterialism: Objects and Social Theory*. Cambridge: Polity, 2016.

Hartley, Daniel. "Against the Anthropocene." *Salvage*, 31 Aug. 2015. Web.

Heise, Ursula. "Science and Ecocriticism." *American Book Review* 18. 5 (1997): 4–6.

Heise, Ursula. "Eco-Narratives." *Routledge Encyclopedia of Narrative Theory*. Edited by David Herman, Manfred Jahn, and Marie-Laure Ryan. London: Routledge, 2005. 129–130.

Heise, Ursula. *Sense of Place and Sense of Planet: The Environmental Imagination of the Global*. Oxford: Oxford University Press, 2008.

Heise, Ursula. "Comparative Literature and the Environmental Humanities." *ACLA State of the Discipline Report*, 2014. Web.

Heise, Ursula. "What's the Matter with Dystopia?" *Public Books*, 2 Jan. 2015. Web.

Heise, Ursula. *Imagining Extinction: The Cultural Meanings of Endangered Species*. Chicago, IL: University of Chicago Press, 2016.

Heise, Ursula. "Climate Stories: Review of Amitav Ghosh's *The Great Derangement*." *Boundary 2*, 19 Feb. 2018. Web.

Helmreich, Stefan. *Alien Ocean: Anthropological Voyages in Microbial Seas*. Berkeley: University of California Press, 2009.

Helmreich, Stefan. "What Was Life? Answers from Three Limit Biologies." *Critical Inquiry* 37. 4 (2011): 671–696.
Hicks, Heather. *The Post-Apocalyptic Novel in the Twenty-First Century: Modernity beyond Salvage*. New York: Palgrave Macmillan, 2016.
Higgins, David. *British Romanticism, Climate Change, and the Anthropocene: Writing Tambora*. Cham: Palgrave Macmillan, 2017.
Hitchcock, Peter. "Oil in an American Imaginary." *New Formations* 69 (2010): 81–97.
Honig, Bonnie. *Public Things: Democracy in Disrepair*. New York: Fordham University Press, 2017.
Horn, Eva. "Überlebensgemeinschaften: Zur Biopolitik der Katastrophe." *Merkur* 773–74 (2013): 992–1004.
Horton, Zach. "Composing a Cosmic View: Three Alternatives for Thinking Scale in the Anthropocene." *Scale in Literature and Culture*. Edited by Michael Tavel Clarke and David Wittenberg. Cham: Palgrave Macmillan, 2017. 35–60.
Hurley, Jessica. "An Apocalypse Is a Relative Thing: An Interview with N. K. Jemisin." *ASAP/Journal* 3. 3 (2018): 467–477.
Ingold, Tim. "Globes and Spheres: The Topology of Environmentalism." *The Perception of the Environment: Essays on Livelihood, Dwelling, and Skill*. London: Routledge, 2000. 209–218.
Iovino, Serenella. "Stories from the Thick of Things: Introducing Material Ecocriticism." *ISLE* 19. 3 (2012): 449–460.
Iovino, Serenella, and Serpil Oppermann. "Introduction: Stories Come to Matter." *Material Ecocriticsm*. Edited by Serenella Iovino and Serpil Oppermann. Bloomington: Indiana University Press, 2014. 1–17.
Irr, Caren. "Climate Fiction in English." *Oxford Research Encyclopedia of Literature*, Feb. 2017. Web.
James, David. "John Burnside's Ecologies of Solace: Regional Environmentalism and the Consolations of Description." *Modern Fiction Studies* 58. 3 (2012): 600–615.
James, Erin. *The Storyworld Accord: Econarratology and Postcolonial Narratives*. Lincoln: University of Nebraska Press, 2015.
Jemisin, N.K. *The Fifth Season*. New York: Orbit, 2016.
Johnson, Mat. *Pym*. New York: Spiegel & Grau, 2012.
Johns-Putra, Adeline. "The Rest Is Silence: Postmodern and Postcolonial Possibilities in Climate Change Fiction." *Studies in the Novel* 50. 1 (2018): 26–42.
Kaplan, E. Ann. *Climate Trauma: Foreseeing the Future in Dystopian Film and Fiction*. New Brunswick: Rutgers University Press, 2015.
Keller, Lynn. "The Ecopoetics of Hyperobjects: Evelyn Reilly's Styrofoam." *ISLE* 22. 4 (2015): 846–871.
Kincaid, Jamaica. *A Small Place*. New York: Farrar, Straus and Giroux, 2000.
King, Thomas. *The Back of the Turtle*. Toronto: Harper Perennial, 2014.
Knapp, Steven, and Walter Benn Michaels. "Against Theory." *Critical Inquiry* 8. 4 (1982): 723–742.

Kolbert, Elizabeth. *The Sixth Extinction: An Unnatural History*. New York: Henry Holt, 2014.

Konstantinou, Lee. "A Theory of Here." *The Account* 4 (2015). Web.

Kramnick, Jonathan, and Anahid Nersessian. "Form and Explanation." *Critical Inquiry* 43. 3 (2017): 650–669.

Kross, Karin. "William Gibson on Urbanism, Science Fiction, and Why *The Peripheral* Weirded Him Out." Tor.com, 29 Oct. 2014. Web.

Kurzweil, Ray. *The Singularity Is Near: When Humans Transcend Biology*. New York: Penguin, 2005.

Larkin, Brian. "The Politics and Poetics of Infrastructure." *Annual Review of Anthropology* 42 (2013): 327–343.

Larkin, Brian. "Promising Forms: The Political Aesthetics of Infrastructure." *The Promise of Infrastructure*. Edited by Hannah Appel, Nikhil Anand, and Akhil Gupta. Durham, NC: Duke University Press, 2018, 175–201.

Latour, Bruno. "On Actor-Network Theory: A Few Clarifications." *Soziale Welt* 47. 4 (1996): 369–381.

Latour, Bruno. "Why Has Critique Run out of Steam? From Matters of Fact to Matters of Concern." *Critical Inquiry* 30. 2 (2004): 225–248.

Latour, Bruno. *Facing Gaia: Eight Lectures on the New Climatic Regime*. London: Polity, 2017.

Lazier, Benjamin. "Earthrise, or, The Globalization of the World Picture." *American Historical Review* 116. 3 (2011): 602–630.

Leane, Elizabeth. *South Pole: Nature and Culture*. London: Reaktion, 2016.

Ledgard, J.M. *Submergence*. London: Vintage, 2012.

Ledgard, J.M. *Terra Firma Triptych: When Robots Fly*. New York: FSG Originals, 2015.

Lee, Sue-Im. "'We are not the world': Global Village, Universalism, and Karen Tei Yamashita's *Tropic of Orange*." *Modern Fiction Studies* 53. 3 (2007): 501–527.

Lehtimäki, Markku. "Natural Environments in Narrative Contexts: Cross-Pollinating Ecocriticism and Narrative Theory." *Storyworlds* 5 (2013): 119–141.

LeMenager, Stephanie. *Living Oil: Petroleum Culture in the American Century*. Oxford: Oxford University Press, 2014.

LeMenager, Stephanie. "Climate Change and the Struggle for Genre." *Anthropocene Reading: Literary History in Geologic Times*. Edited by Tobias Menely and Jesse Oak Taylor. University Park: Pennsylvania State University Press, 2017a. 220–238.

LeMenager, Stephanie. "The Humanities after the Anthropocene." *The Routledge Companion to the Environmental Humanities*. Edited by Ursula Heise, Jon Christensen, and Michelle Niemann. London: Routledge, 2017b. 473–481.

Lemke, Thomas. "Materialism without Matter: The Recurrence of Subjectivism in Object-Oriented Ontology." *Distinktion* 18. 2 (2017): 133–152.

Lemke, Thomas. "An Alternative Model of Politics? Prospects and Problems of Jane Bennett's Vital Materialism." *Theory, Culture and Society* 35. 6 (2018): 31–54.

Lerner, Ben. *10:04*. London: Granta, 2014.

Levine, Caroline. "Infrastructuralism, or, The Tempo of Institutions." *On Periodization: Selected Essays from the English Institute*. Edited by Virginia Jackson. Cambridge, MA: English Institute, 2010. 53–96. Web.

Levine, Caroline. *Forms: Whole, Rhythm, Hierarchy, Network*. Princeton, NJ: Princeton University Press, 2015.

Levinson, Marjorie. "What Is New Formalism?" *PMLA* 122. 2 (2007): 558–569.

Lewis, Simon, and Mark Maslin. "Defining the Anthropocene." *Nature* 519 (2015): 171–180.

Lorimer, Jamie. *Wildlife in the Anthropocene: Conservation after Nature*. Minneapolis: University of Minnesota Press, 2015.

Lovecraft, H.P. "Supernatural Horror in Literature." *H. P. Lovecraft: The Complete Fiction*. New York: Barnes & Noble, 2011, 1041–1098.

Lovelock, James. *Gaia: A New Look at Life on Earth*. Oxford: Oxford University Press, 2000.

Luckhurst, Roger. "The Weird: A Dis/Orientation." *Textual Practice* 31.6 (2017): 1041–1061.

Lyotard, Jean-François. *The Inhuman: Reflections on Time*. Stanford, CA: Stanford University Press, 1992.

Malm, Andreas. *Fossil Capital: The Rise of Steam Power and the Roots of Global Warming*. London: Verso Books, 2016

Malm, Andreas, and Ralf Holborg. "The Geology of Mankind? A Critique of the Anthropocene Narrative." *The Anthropocene Review* 1. 1 (2014): 62–69.

Mandel, Emily St. John. *Station Eleven*. London: Picador, 2014.

Mann, Charles. *1493: Uncovering the New World Columbus Created*. New York: Vintage, 2011.

Marcus, Sharon, Heather Love, and Stephen Best. "Building a Better Description." *Representations* 135 (2016): 1–21.

Marland, Pippa. "Ecocriticism." *Literature Compass* 10–11 (2013): 846–868.

Marris, Emma. *Rambunctious Garden: Saving Nature in a Post-Wild World*. New York: Bloomsbury, 2011.

Marris, Emma. "Handle with Care." *Orion Magazine* 34. 3 (2015). Web.

Marshall, Kate. "What Are the Novels of the Anthropocene? American Fiction in Geological Time." *American Literary History* 27. 3 (2015): 523–538.

Martel, Yann. *Life of Pi*. New York: Harcourt, 2001.

Martin, Theodore. *Contemporary Drift: Genre, Historicism, and the Problem of the Present*. New York: Columbia University Press, 2017.

Martinez-Alier, Joan. *The Environmentalism of the Poor: A Study of Ecological Conflicts and Valuation*. New York: Edward Elgar, 2002.

Masco, Joseph. "The Age of Fallout." *History of the Present* 5. 2 (2015): 137–168.

Mattern, Shannon. "Ear to the Wire: Listening to Historic Urban Infrastructures." *Amodern* 2 (2013a). Web.

Mattern, Shannon. "Infrastructural Tourism." *Places Journal*, July 2013b. Web.

McCarthy, Tom. "The Death of Writing: If James Joyce Were Alive Today He'd Be Working for Google." *The Guardian*, 7 Mar. 2015a. Web.

McCarthy, Tom. *Satin Island*. New York: Knopf, 2015b.

McGuire, Richard. *Here*. New York: Pantheon Books, 2014.
McGurl, Mark. "The Posthuman Comedy." *Critical Inquiry* 38. 3 (2012): 533–553.
McGurl, Mark. "Gigantic Realism: The Rise of the Novel and the Comedy of Scale." *Critical Inquiry* 43. 2 (2017): 403–430.
McKibben, Bill. *Eaarth: Making a Life on a Tough New Planet*. New York: Times Books, 2010.
Meillassoux, Quentin. *After Finitude: An Essay on the Necessity of Contingency*. London: Bloomsbury, 2006.
Mentz, Steve. *Break Up the Anthropocene*. Minneapolis: University of Minnesota Press, 2019.
Mirzoeff, Nicholas. "Visualizing the Anthropocene." *Public Culture* 26. 2 (2014): 213–232.
Mirzoeff, Nicholas. *How to See the World*. New York: Penguin, 2015.
Mitchell, David. *Cloud Atlas*. London: Sceptre, 2004.
Moore, Jason. "The Capitalocene, Part I: On the Nature and Origins of Our Ecological Crisis." *Journal of Peasant Studies* 44. 3 (2017): 594–630.
Moretti, Franco. "Serious Century." *The Novel, Volume I: History, Geography, and Culture*. Edited by Franco Moretti. Princeton, NJ: Princeton University Press, 2007, 364–400.
Morton, Timothy. "Here Comes Everything: The Promise of Object-Oriented Ontology." *Qui Parle* 19. 2 (2011): 163–190.
Muñoz, José Esteban. *Cruising Utopia: The Then and There of Queer Futurity*. New York: New York University, 2009.
Musk, Elon. "Making Humans a Multi-Planetary Species." *New Space* 5. 2 (2017): 46–61.
Nancy, Jean-Luc. *The Creation of the World, or, Globalization*. Albany, NY: SUNY Press, 2007.
Newitz, Annalee. "William Gibson on the Apocalypse, America, and *The Peripheral*'s Ending." *Gizmodo*, 10 Nov. 2014. Web.
Neyrat, Frédéric. "Eccentric Earth." *Diacritics* 45. 3 (2017): 4–21.
Ngai, Sianne. *Ugly Feelings*. Cambridge, MA: Harvard University Press, 2007.
Nixon, Rob. *Slow Violence and the Environmentalism of the Poor*. Cambridge, MA: Harvard University Press, 2011.
Nixon, Rob. "The Great Acceleration and the Great Divergence: Vulnerability in the Anthropocene." *MLA Profession*, Mar. 2014. Web.
Oliver, Kelly. *Earth and World: Philosophy after the Apollo Missions*. New York: Columbia Univeristy Press, 2015.
Oppermann, Serpil. "From Ecological Postmodernism to Material Ecocriticism: Creative Materiality and Narrative Agency." *Material Ecocriticsm*. Edited by Serenella Iovino and Serpil Oppermann. Bloomington: Indiana University Press, 2014. 21–36.
Oppermann, Serpil. "From Posthumanism to Posthuman Ecocriticism." *Relations* 4. 1 (2016): 23–37.
Oppermann, Serpil. "The Scale of the Anthropocene: Material Ecocritical Reflections." *Mosaic* 51. 3 (2018): 1–17.

Oreskes, Naomi, and Erik M. Conway. *Merchants of Doubt: How a Handful of Scientists Obscured the Truth on Issues from Tobacco Smoke to Global Warming*. London: Bloomsbury, 2010.

Oreskes, Naomi, and Erik M. Conway. *The Collapse of Western Civilization: A View from the Future*. New York: Columbia University Press, 2014.

Parikka, Jussi. *A Geology of Media*. Minneapolis: University of Minnesota Press, 2015.

Petrocultures Research Group. *After Oil*. Edmonton: Petrocultures Research Group, 2016.

Pinkus, Karen. *Fuel: A Speculative Dictionary*. Minneapolis: University of Minnesota Press, 2016.

Plumwood, Val. *Environmental Culture: The Ecological Crisis of Reason*. London: Routledge, 2002.

Poe, Edgar Allan. *The Narrative of Arthur Gordon Pym of Nantucket*. New York: Penguin, 1999.

Povinelli, Elizabeth. *Geontologies: A Requiem to Late Liberalism*. Durham, NC: Duke University Press, 2016.

Purdy, Jedediah. *This Land is Our Land: The Struggle for a New Commonwealth*. Princeton, NJ: Princeton University Press, 2019.

Reed, Christina. "Plastic Age: How It's Reshaping Rocks, Oceans, and Life." *New Scientist* 3006, 28 Jan. 2015. Web.

Reilly, Evelyn. *Styrofoam*. New York: Roof Books, 2009.

Rekret, Paul. "A Critique of New Materialism: Ethics and Ontology." *Subjectivity* 9. 3 (2016): 225–225.

Robbins, Bruce. "The Smell of Infrastructure: Notes toward an Archive." *Boundary 2* 34. 1 (2007): 25–33.

Robbins, Bruce. *The Beneficiary*. Durham, NC: Duke University Press, 2017.

Rody, Caroline. *The Interethnic Imagination: Roots and Passages in Contemporary Asian American Fiction*. Oxford: Oxford University Press, 2009.

Ronda, Margaret. "Anthropogenic Poetics." *Minnesota Review* 83 (2014): 102–111.

Rothstein, Adam. "How to See Infrastructure: A Guide for Seven Billion Primates." Rhizome.org, 2 July 2015. Web.

Rubenstein, Michael, Bruce Robbins, and Sophia Beal. "Infrastructuralism: An Introduction." *Modern Fiction Studies* 61. 4 (2015): 575–586.

Ruddiman, William. "The Anthropogenic Greenhouse Era Began Thousands of Years Ago." *Climate Change* 61 (2003): 261–293.

Scheffler, Samuel. *Death and the Afterlife*. Oxford: Oxford University Press, 2016.

Scranton, Roy. *Learning to Die in the Anthropocene: Reflections on the End of a Civilization*. San Francisco, CA: City Lights Books, 2015.

Scranton, Roy. *We're Doomed. Now What? Essays on War and Climate Change*. New York: Soho Press, 2018.

Sefton-Rowston, Adelle. "Sovereignty as a State of Craziness: Empowering Female Indigenous Psychologies in Australian 'Reconciliatory Literature'." *Hypatia* 32. 3 (2017): 644–659.

Seltzer, Mark. *The Official World*. Durham, NC: Duke University Press, 2016.

Shklovsky, Viktor. "Art as Technique." *Russian Formalist Criticism: Four Essays*. Edited by Lee T. Lemon and Marion J. Reis. Lincoln: Nebraska University Press, 1965. 3–24.
Siskind, Mariano. "Captain Cook and the Discovery of Antarctica's Modern Specificity: Towards a Critique of Globalization." *Comparative Literature Studies* 42. 1 (2005): 1–23.
Slovic, Scott. "The Third Wave of Ecocriticism: North American Reflections on the Current Phase of the Discipline." *Ecozon@* 1. 1 (2010): 4–10.
Slovic, Scott. "Editor's Note." *ISLE* 19. 4 (2012): 619–621.
Smith, Rachel Greenwald. *Affect and American Literature in the Age of Neoliberalism*. Cambridge: Cambridge University Press, 2015.
Soni, Vivasvan. "Energy." *Fueling Culture: 101 Words for Energy and Environment*. Edited by Imre Szeman, Jennifer Wenzel, and Patricia Yaeger. New York: Fordham University Press, 2017. 132–135.
Spahr, Juliana. *Well Then There Now*. Boston, MA: Black Sparrow Books, 2011.
Steinberg, Philip. "Of Other Seas: Metaphors and Materialities in Maritime Regions." *Atlantic Studies* 10. 2 (2013): 156–169.
Streeby, Shelley. *Imagining the Future of Climate Change: World-Making through Science Fiction and Activism*. Oakland: University of California Press, 2018.
Sze, Julie. "'Not by politics alone': Gender and Environmental Justice in Karen Tei Yamashita's *Tropic of Orange*." *Bucknell Review* 44. 1 (2000): 29–42.
Szeman, Imre. "Globalization." *The Johns Hopkins Guide to Literary Theory and Criticism*. Edited by Michael Groden, Martin Kreiswirth, and Imre Szeman. Baltimore, MA: Johns Hopkins University Press, 2005. 458–465.
Szeman, Imre. "Literature and Energy Futures." *PMLA* 126. 2 (2011): 323–325.
Szeman, Imre. *On Petrocultures: Globalization, Culture, and Energy*. Morgantown: West Virginia University Press, 2019.
Szerszynski, Bronislaw. "The End of the End of Nature: The Anthropocene and the Fate of the Human." *Oxford Literary Review* 34. 2 (2012): 165–184.
Szerszynski, Bronislaw. "How the Earth Remembers and Forgets." *Political Geology: Active Stratigraphies and the Making of Life*. Edited by Adam Bobbette and Amy Donovan. Cham: Palgrave Macmillan, 2019. 219–236.
Takolander, Maria. "Theorising Irony and Trauma in Magical Realism: Junot Díaz's *The Brief Wondrous Life of Oscar Wao* and Alexis Wright's *The Swan Book*." *Ariel* 47. 3 (2016): 95–122.
Tate, Andrew. *Apocalyptic Fiction*. London: Bloomsbury, 2017.
Taylor, Jesse Oak. *The Sky of Our Manufacture: The London Fog in British Fiction from Dickens to Woolf*. Charlottesville: University of Virginia Press, 2016.
Taylor, Matthew. "At Lands's End: Novel Spaces and the Limits of Planetarity." *Novel* 49. 1 (2016): 115–138.
Thacker, Eugene. *After Life*. Chicago, IL: University of Chicago Press, 2010.
Thacker, Eugene. *In the Dust of This Planet: Horror of Philosophy Vol. 1*. Winchester: Zero Books, 2011.
Thill, Brian. *Waste*. London: Bloomsbury, 2015.

Trexler, Adam. *Anthropocene Fictions: The Novel in a Time of Climate Change.* Charlottesville: University of Virginia Press, 2015.

VanderMeer, Jeff. *Annihilation.* New York: Farrar, Straus and Giroux, 2014a.

VanderMeer, Jeff. *Authority.* New York: Farrar, Straus and Giroux, 2014b.

van Dooren, Thom. *Flight Ways: Life and Loss at the Edge of Extinction.* New York: Columbia University Press, 2014.

Wallace-Wells, David. *The Uninhabitable Earth: Life after Warming.* New York: Tim Duggan Books, 2019.

Weik von Mossner, Alexa. "Environmental Narrative, Embodiment, and Emotion." *Handbook of Ecocriticism and Cultural Ecology.* Edited by Hubert Zapf. Berlin: de Gruyter, 2016. 534–550.

Weik von Mossner, Alexa. *Affective Ecologies: Empathy, Emotion, and Environmental Narrative.* Columbus: Ohio State University Press, 2017.

Weisman, Alan. *The World without Us.* New York: Picador, 2008.

Wenzel, Jennifer. "Planet vs. Globe." *English Language Notes* 52. 1 (2014): 19–30.

Wenzel, Jennifer. "Introduction." *Fueling Culture: 101 Words for Energy and Environment.* Edited by Imre Szeman, Jennifer Wenzel, and Patricia Yaeger. New York: Fordham University Press, 2017. 1–16.

Whyte, Kyle Powys. "Our Ancestors' Dystopia Now: Indigenous Conservation and the Anthropocene." *The Routledge Companion to the Environmental Humanities.* Edited by Ursula Heise, Jon Christensen, and Michelle Niemann. London: Routledge, 2017. 206–215.

Williams, Raymond. *Marxism and Literature.* Oxford: Oxford University Press, 1977.

Wilson, Eric. "Polar Apocalypse in Coleridge and Poe." *Wordsworth Circle* 35. 1 (2004): 37–44.

Winkiel, Laura. "Introduction." *English Language Notes* 57. 1 (2019): 1–10.

Wolfe, Cary. *What Is Posthumanism?* Minneapolis: University of Minnesota Press, 2010.

Wood, Gillen D'Arcy. *Tambora: The Eruption That Changed the World.* Princeton, NJ: Princeton University Press, 2014.

Woods, Derek. "Scale Critique for the Anthropocene." *Minnesota Review* 83 (2014): 133–142.

Woods, Derek. "Scale Variance and the Concept of Matter." *The New Politics of Materialism: History, Philosophy, Science.* Edited by Sarah Ellenzweig and John H. Zammito. London: Routledge, 2017. 200–224.

Woods, Derek. Review of *The Anthropocene as a Threshold Concept*, by Timothy Clark. *Configurations* 26. 4 (2018): 502–504.

Wright, Alexis. *The Swan Book.* London: Little, Brown, 2016.

Yaeger, Patricia. "Introduction: Dreaming of Infrastructure." *PMLA* 122. 1 (2007): 9–26.

Yaeger, Patricia. "Literature in the Ages of Wood, Tallow, Coal, Whale Oil, Gasoline, Atomic Power, and Other Energy Sources." *PMLA* 126. 2 (2011): 305–310.

Yamashita, Karen Tei. *Tropic of Orange.* Minneapolis: Coffee House Press, 2017.

Zylinska, Joanna. "Photography after Extinction." *After Extinction*. Edited by Richard Grusin. Minneapolis: University of Minnesota Press, 2018. 51–70.

Zalasiewicz, Jan. *The Earth after Us: What Legacy Will Humans Leave in the Rocks?* Oxford: Oxford University Press, 2009.

Zalasiewicz, Jan, et al. "The Working Group on the Anthropocene: Summary of Evidence and Interim Recommendations." *Anthropocene* 19 (2017): 55–60.

Index

For Product Safety Concerns and Information please contact our EU
representative GPSR@taylorandfrancis.com
Taylor & Francis Verlag GmbH, Kaufingerstraße 24, 80331 München, Germany

www.ingramcontent.com/pod-product-compliance
Ingram Content Group UK Ltd.
Pitfield, Milton Keynes, MK11 3LW, UK
UKHW020130200926
475702UK00005B/682
* 9 7 8 1 1 3 8 5 4 3 7 1 3 *